T0135534

Silesian University of Technology
Faculty of Automatic Control, Electronics
and Computer Science

Examination and simulation of new magnetic materials for the possible application in memory cells

Andrea Ehrmann

PhD dissertation written under guidance of
Dr. hab. Tomasz Błachowicz
Prof. at Silesian University of Technology

Gliwice (2013)

Bibliografische Information der Deutschen Nationalbibliothek

Die Deutsche Nationalbibliothek verzeichnet diese Publikation in der
Deutschen Nationalbibliografie; detaillierte bibliografische Daten sind
im Internet über http://dnb.d-nb.de abrufbar.

Coverbild: REM-Aufnahme, ACCESS e. V.

ISBN 978-3-8325-3772-2

Logos Verlag Berlin GmbH
Comeniushof, Gubener Str. 47,
10243 Berlin
Tel.: +49 (0)30 42 85 10 90
Fax: +49 (0)30 42 85 10 92
INTERNET: http://www.logos-verlag.de

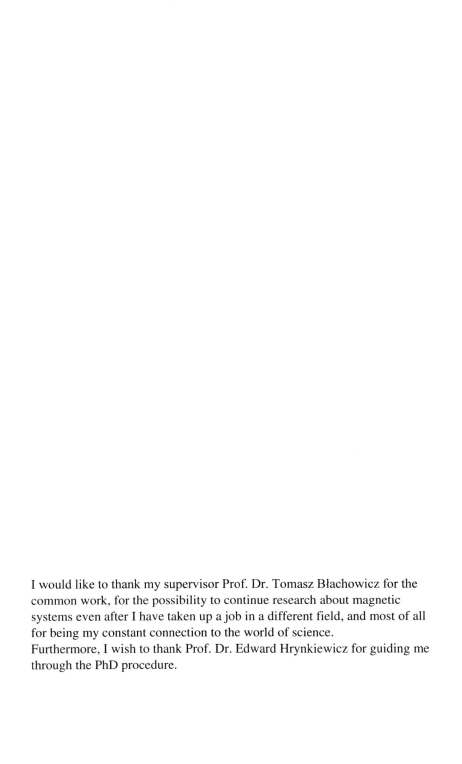

I would like to thank my supervisor Prof. Dr. Tomasz Błachowicz for the common work, for the possibility to continue research about magnetic systems even after I have taken up a job in a different field, and most of all for being my constant connection to the world of science.

Furthermore, I wish to thank Prof. Dr. Edward Hrynkiewicz for guiding me through the PhD procedure.

Contents

1. Introduction 7
2. Theoretical foundations 10
   2.1. Motivation 10
   2.2. History of storage media 11
   2.3. Ideas for future magnetic storage media 19
   2.4. Technical interest in systems with novel anisotropies for data storage media 21
3. Experimental results 23
   3.1. MOKE setup 23
   3.2. Other methods to measure magnetization 25
   3.3. Preparation of magnetic samples with textile methods 26
   3.4. MOKE experiments on thin layer samples (Co/CoO) 30
   3.5. Comparison with results of Magpar simulations 32
   3.6. Comparison with results of MathCad simulations 34
   3.7. MOKE experiments on nano-cylinders 38
   3.8. MOKE experiments on textile-based samples with uneven surface ..38
4. Simulations with Magpar 40
   4.1. MagPar – a short overview of the program 40
   4.2. Low-dimensional half-ball systems with shape modifications 41
      4.2.a. Hysteresis loops 42
      4.2.b. Reversal dynamics 50
      4.2.c. Vortex core precession and M rotation 54
      4.2.d. Influence of the Gilbert damping constant α 57
      4.2.e. Influence of the field sweeping speed 58
      4.2.f. Magnetization reversal mechanisms in M_x-M_y-graphs 61
      4.2.g. Shape dependence of magnetization reversal 63
   4.3. 4-fold wire systems 73
      4.3.a. Hysteresis loops with intermediate states 74
      4.3.b. Angular dependence of the coercive fields – 2x2 wires 78
      4.3.c. Influence of dimensions on magnetization reversal processes 86
      4.3.d. Magnetization reversal mechanisms in M_x-M_y-graphs 92
      4.3.e. Influence of wire connections 93
      4.3.f. Special corner solutions..95
      4.3.g. Magnetization reversal in fourfold Co wire system 98
   4.4. 2-fold wire systems 102
      4.4.a. Angular dependence of the coercive fields – parallel wires 102
      4.4.b. Comparison with wire sample with extended ends 112
   4.5. 3-fold wire system 116
   4.6. 6-fold wire system 119
5. Outlook 124
   5.1. Systems with novel anisotropies 124

    5.2.   Proposals for technological solution 130

6. Conclusion 135

Literature...138

Appendix I: Anisotropies and internal magnetic fields in FM systems 155

Appendix II: Description of MathCad program used in this thesis..............159

Appendix III: Comparison of theoretical and experimental results 163

1. Introduction

Magnetic moments in a crystal are influenced by magnetic anisotropies, which define magnetically easy (favored) and hard directions relative to the crystal lattice. These anisotropies are often caused by crystal symmetries, such as the form anisotropy, interface anisotropy, or magneto-crystalline anisotropy. Additionally, by coupling a ferromagnetic and an antiferromagnetic material, the so-called exchange bias can arise, a unidirectional anisotropy which was discovered in 1956 [Mei56]. The exchange bias can lead to a horizontal shift of the hysteresis loop and to a change of the shape of the loop, often broadening it and sometimes causing steps or other unusual features.

In nanostructured materials, the form anisotropy normally plays the most important role. Moreover, examinations of the precessional switching, e.g., a very fast method of changing the magnetization direction which could be utilized for MRAMs (Magnetic Random Access Memory) and similar devices, have shown that even in particles of lateral dimensions 7 μm x 20 μm the magnetization at the edges remains pinned, resulting in a back-switching of the magnetization in the middle of the sample [Hie03]. These findings are the motivation to examine nano- and microstructured samples in simulation and experiment in this thesis.

Due to the potential utilization in various applications, the reversal mechanisms and dynamics of nanostructured magnetic systems have been intensively studied during the last decade. Compared with common ultra-thin single layers and multilayer systems, magnetic nanostructures can give rise not only to novel static and dynamic behavior but also to new anisotropies, contributing to the development of spintronic devices.

Since possible applications include data storage media, examinations of nanostructured systems often aim at decreasing the pattern size, in order to enhance the possible information density in a given area. In this thesis, another approach is chosen: Intermediate magnetic states occurring during magnetization reversal, which are stable at zero external field, with a remanence different from the value in the complete hysteresis loop, can lead to quaternary or higher-order multilevel magnetic storage media. In this way, the storage density can be enhanced without decreasing the size of the magnetic nanosystems.

While such intermediate states have already been recognized in exchange bias systems such as Fe/MnF_2, searching for stable states in pure ferromagnetic systems is of great interest for future developments in data storage systems. This thesis describes different nanostructured systems in which such additional stable states can be found. The intermediate states are correlated with different magnetic states and different reversal dynamics. The anisotropies of such nanosystems are examined and compared with theoretical and experimental findings of thin layer systems. Recommendations are given for shape and material of future nanostructured systems for data storage media.

In Chapter 2, the theoretical foundations of this work are described after explaining the motivation for this thesis in detail. An overview of the technical state-of-the-art and recent research results according to data storage media is given, leading to ideas for future magnetic storage media. The technical interest in systems with novel anisotropies is explained.

In Chapter 3, the experimental setup for measurements of the Magneto-Optical Kerr-Effect (MOKE) is described, followed by a short introduction into other possibilities to measure the magnetization of a sample. Experimental results of MOKE measurements on own thin-layer samples consisting of Co/CoO bilayers are provided. Details of the sample characterization are given to understand different anisotropies in these samples. As an example of MOKE on nanostructured samples, measurements on nano-cylinders with asymmetrical holes are shown. Additionally, some attempts of MOKE measurements on textile-based samples with typically very uneven surfaces are described. Finally, experimental results are compared with simulations.

In Chapter 4, the Parallel Finite Element Micromagnetics Package MAGPAR [Sch03] is described as well as the meshing process which is performed by the GiD personal pre and post processor software. Results of simulations with MAGPAR are shown. Firstly, low-dimensional half-ball systems are simulated. Hysteresis loops and the reversal dynamics are depicted depending on several shape modifications. Additionally, the vortex core precession is examined, and the results are put in context with literature findings. The next step of the simulations concentrates on 4-fold wire systems. In these systems, stable intermediate magnetization states have been found, similar to some exchange bias systems. Different magnetization reversal processes have been observed which are correlated with the existence or absence of an intermediate state. For a large range of wire lengths and diameters, the influence of the system dimensions on the magnetization reversal processes is examined. In 6-fold wire systems, similar intermediate states can be observed in the hysteresis loops, again related to certain magnetization reversal processes. It can be shown that the corner shape strongly influences the system's magnetic properties.

In Chapter 5 an outlook on future research is given.

Finally, in Chapter 6, the theoretical and experimental results are concluded.

The appendices give the mathematical base of anisotropy modeling as well as a description of the MathCad program written for this thesis and show an overview of the comparison between theoretical and experimental results.

2. Theoretical foundations

In this chapter, after a short motivation for this work, an overview is given on former and recent storage media. Different magnetic anisotropies and typical mathematical models found in literature are discussed, alongside with novel ideas for future magnetic storage media. To underline the importance of understanding common and novel anisotropies, the ideas of quaternary storage media and similar technical applications are explained.

2.1. Motivation

During the past years, several own simulations have supported the idea of a novel magnetic storage memory system – a system with two bits in one pattern, correlated to four stable states at vanishing external magnetic field, thus doubling the data storage density without changing the number of "places" on a hard disk which are read / written.

Similar to bit-patterned media using dots, as usual in recent trials by Toshiba [Tos10] (Fig. 2.1, left panel), the idea of this thesis is utilization of different structures as dual-bit-patterned media (right panel).

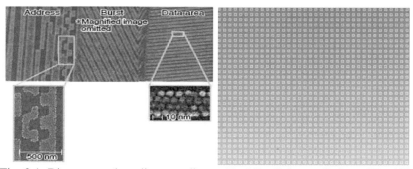

Fig. 2.1: Bit-patterned media according to Toshiba (left panel), from [Tos10], and own structure for dual-bit-patterned media (right panel).

It should be mentioned that, while doubling the number of stable states in a hysteresis loop from 2 to 4 doubles the number of bits which can be stored in the respective magnetic particle, this process changes quantitatively for the next numbers: 3 bits can store 8 different combinations of 1 and 0, thus 8 stable states (not 6) would be necessary to store 3 bits of information.

Former research has been published in the following papers:

   I.   T. Blachowicz, A. Ehrmann neé Tillmanns, P. Steblinski, L. Pawela: Magnetization reversal in magnetic half-balls influenced by shape perturbations, J. Appl. Phys. **108**, 123906 (2010)

II. Ehrmann, T. Błachowicz: Adjusting exchange bias and coercivity of magnetic layered systems with varying anisotropies, J. Appl. Phys. **109**, 083923 (2011)

III. T. Blachowicz, A. Ehrmann: Fourfold nanosystems for quaternary storage devices, J. Appl. Phys. **110**, 073911 (2011)

IV. Ehrmann, T. Błachowicz, P. Steblinski, M. O. Weber: Neue Anisotropien - von der Grundlagenforschung zu optimierten magnetischen Speichermedien, in A. Brenke (Ed.): ASIM-Konferenz STS/GMMS 2011, Proceedings, Shaker Verlag (2011)

V. T. Blachowicz, A. Ehrmann: Anatomy of Demagnetizing and Exchange Fields in Magnetic Nanodots Influenced by 3D Shape Modifications, arXiv:1207.4673v1 (2012)

VI. T. Blachowicz, A. Ehrmann: Six-state, three-level, six-fold ferromagnetic wire system, J. Magn. Magn. Mat. **331**, 21-23 (2013)

VII. T. Blachowicz, A. Ehrmann, P. Steblinski, J. Palka: Directional-dependent coercivities and magnetization reversal mechanisms in fourfold ferromagnetic systems of varying sizes, J. Appl. Phys. **113**, 013901 (2013)

The contents of these articles are partly included in this thesis.

2.2. History of storage media

Recording of *analog* audio signals on ferromagnetic wires and tapes belongs to the first utilizations of magnetic storage media [Lem2003]. In these systems, developed by Poulsen and O'Neill, respectively, the steel wire / magnetically coated tape was rapidly drawn (with 610 mm/s) over the recording head which magnetized short sections of the medium with regard to the electrical audio signal reaching the recording head at the respective time. Signal intensity and also polarity defined the magnetization in the medium. Reading occurred when performing the same process without an electrical field at the reading head, resulting in an induced electrical signal due to the magnetization in the medium. While first trials used transverse magnetization, later developments changed to longitudinal magnetization which was found to be more stable against a twist in the wire.

While the principle of recording audio or video signals on magnetic tapes has been used for long periods in form of compact cassettes and video tapes, data recording, i.e. storage of *digital* values, is still being used due to the reduced cost per bit, compared with hard disks [Ele10], and the quite long lifetimes of theoretically about 10-30 years. Especially for backups, magnetic tape recording can be used, since in this case it is not necessary to switch rapidly between different recording positions.

The most often used standard used for magnetic tape storage systems is LTO (Linear Tape-Open), initiated by IBM, Seagate, and Hewlett-Packard. Recently, the maximum speed of cartridges using this standard is 160 MB/s [Ult13], the linear density approximately 15 kbits/mm [Qual13], a value which has not been changed between the last two generations [Qual13a]. However, areal densities of about 30 Gb/in^2 have already been shown, according to a tape cartridge capacity of 35 terabytes [Che10]. A theoretical work has pointed out that the areal density could even be increased to 100 Gb/in^2, corresponding with more than 100 TB on one tape cartridge [Arg08].

Other techniques than the described linear method are the linear serpentine technique (with more tracks than heads), the transverse scan writing short data tracks perpendicular to the tape length, and a mixture of both, leading to helical or arc-like tracks.

Other standards are DLT (Digital Linear Tape), developed by Digital Equipment Corporation (DEC), or DDS (Digital Data Storage), both currently having lower capacity and lower data transfer speed.

Similar to recent hard disks, drum memory works in a rotational mode, with a row of read/write heads oriented along the drum axis, with one head per track [IBM55]. Invented in 1932 by Tauschek, the system could store some ten kilobytes and was exchanged by semiconductor memory in the 1970s.

The magnetic-core memory belongs to the random access memories (RAM), i.e. direct access memories in which unpredictable positions can be accessed. The "cores" are small ferromagnetic rings, normally produced from ferrite, each representing one bit, which can be magnetized either clockwise or counterclockwise [Wri51]. The round shape has the advantage of almost no stray fields, allowing the cores to be closely packed. Additionally, neighboring cores are normally oriented 90° to each other, which further diminishes the influence of stray fields [Sel77].

The cores are mounted on a rectangular mesh of fine wires in x- and y-direction, each core inclosing one crossing point of the wires. A third set of wires, the inhibit / sense lines, is oriented at ±45° to the x- and y-wires, passing through the cores [Ols59].

To write a 1 bit, starting from the state 0, a current is applied to the x- and y-wires crossing at the respective intersection point, leading to a change in the magnetic polarity of the core. Only the superposed currents of both lines suffice to change the magnetization in the core, the other cores which surround only one of the two wires are not influenced.

To write a 0 bit, the identical current is applied to the inhibit line, resulting in a reduced net current which is not sufficient to change the magnetic polarity.

Reading works with currents in opposite direction. For a core in the 0 state, nothing happens. If the core was in the 1 state before, the change of the magnetic polarity induces a voltage pulse in the sense line. Since reading changes the state of the core, in case of a 1 before reading, this 1 has to be re-written after reading [Jon76].

Thin-film memory uses fine permalloy dots and printed circuits instead of ferrite cores on wires, leading to fast access times in the order of 670 ns [Nav78].

In twistor memory, developed in the late 1960s, magnetic tape was wrapped around the crossing point of x- and y-wire, with a third dimension added in the form of solenoids. In this way, a complete row of data could be processed at the same time [Mei65]. However, production of the 3D memory became too time-consuming and thus too expensive, compared with newer memories.

In 1956, the first hard disk drive (HDD) was developed [IBM56]. HDDs consist of one or more rotating disks, firstly with a diameter of 14 inches, nowadays reduced to 3.5 inches, 2.5 inches or even smaller diameters. The circular disks, the so-called platters, consist of a non-magnetic material, coated by a thin ferromagnetic layer (usually 10-20 nm) and an additional outer protection layer [Hit10]. One HDD can consist of several platters. Opposite to drum or tape memories, normally only one read/write head is used on each side of each platter (e.g. [Esc13]). All read/write heads are moved by a common arm.

Hard disks nowadays have capacities of some terabytes, after some megabytes in the first models [IBM56]. Access times could be reduced from more than 100 ms to a few milliseconds, with spin speeds around 15,000 rounds per minute [Wes13]. Similarly, the areal data densities could be roughly doubled in every 18 months, similar to the doubling of the number of transistors on a chip, which is described by Moore's law [Moo65]. However, this value has not been reached during the last few years. The data transfer rate is recently in the order of 600 Mbit/s [Sea13].

The read/write head of a hard disk can work with different techniques which have been developed further during the past decades.

The very first heads worked similarly to tape recorder heads. A fine coil was wrapped around a ferrite ring with a small slit. When a current was flowing through the coil, a strong magnetic field was built in the slit which was used to magnetize the platter surface below this area. Reading worked oppositely – the magnetized platter surface induced a current in the coil. By introducing a small piece of metal in the slit, concentrating the field, smaller areas could be addressed. Additional progress could be made by exchanging the 3D geometry by a thin film head which could be structured by photolithography. This technique allowed for production of hard disk drives of some GB storage capacity in the middle of the 1990s [PCG05].

Another big step in the enhancement of HDD capacity was the separation of the read and the write head. Firstly, the read head used the AMR (anisotropic magneto resistance) effect, which results in a change of the electrical resistance due to a magnetic field. In this way, reading smaller magnetic areas was possible [IBM96]. Starting in 1997, GMR (giant magneto resistance) heads were used instead [IBM13, Hit10], consisting of two or more ferromagnetic layers separated by a non-magnetic metal. The dependence of the resistance on the relative magnetization orientation can be in the order of several ten percent, enabling even smaller structures to be read. It can be explained by spin-dependent scattering of the electrons in the material – when all layers are oriented in parallel, either all electrons with spin "up" or all electrons with spin "down" will not experience much scatter in the material; when the layers are oriented in alternating directions, both spin orientations will be scattered, increasing the overall resistance for the sum of electrons with both spin orientations.

The latest heads use TMR (tunnel magneto resistance) elements in which the ferromagnetic layers are separated by non-magnetic isolators. The tunnel current between neighboring layers depends on their respective orientation. The first drives with TMR heads produced by Seagate use miniaturized heating coils which support the writing and reading operations [Che06].

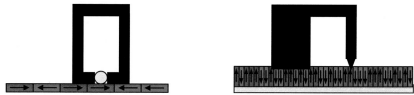

Fig. 2.2: Sketch of difference between standard in-plane "longitudinal" recording (left panel) and perpendicular recording (right panel).

While the read/write heads have been improved during the past decade, an additional change has started in terms of the magnetic material on the platter: Since the storage density for conventional in-plane recording is restricted by the superparamagnetic limit (if the structures are too small, thermal energy becomes sufficient to change the magnetization arbitrarily – the stability is higher for larger regions and larger coercivities), researchers tried to switch to perpendicular recording (for a review, see [Pir07]).

In this geometry, on the one hand, the single bits use less space in the recording layer (cf. Fig. 2.2). On the other hand, materials with higher coercivity can be used, since the magnetic flux in perpendicular writing, passing through the recording layer into the additional soft magnetic layer (marked grey in Fig. 2.2) and back into the broad part of the writing head, flows to a greater extent through the single bits, allowing for materials with higher coercive fields to be magnetized by similar currents in the write head.

First perpendicular recording HDDs have been produced by Toshiba in 2004 [Tos04] and Seagate in 2005 [Smi05].

While hard disk drives are used as re-writable media, magnetic stripe cards are made to store small amounts of data which are never or not often changed. Magnetic iron particles in a flat band, similar to those described above, can carry information, e.g. in credit cards, access cards, ID cards, or the like. Bank cards and other cards with important data use high-coercivity bands, to avoid data loss due to accidental contact with magnets [IBM13a].

Another technique used in the banking industry is magnetic ink which was developed in the 1950s, especially for processing checks. With the check information transformed into numbers written with magnetic ink, which can be identified by a special read head, checks can be processed in an automated way [MIC13].

In the early 1960s, external memory cards, so-called CRAM cards, were produced to offer the possibility of exchanging the data which was to be processed. These cards were handled with vacuum and blowers, allowing for one card to be read/written and afterwards be carried back to the stack of cards, while another card could be brought to the magnetic read/write head [NCR62].

The next generation of external memory is the floppy disk, developed in the late 1960s. Unlike hard disk drives, early floppy disks had one or several "index holes", real holes in the magnetic disk which could be used to detect the starting angle and the rotation speed. Additionally, the housings had – depending on the dimension of the floppy disk – different slot solutions which were used to allow/forbid writing by closing the slot by a tape or leaving it open. The magnetic disk consisted, e.g., from iron oxide, cobalt or barium ferrite, coated on both sides with a nonmagnetic material, such as Teflon, to reduce friction in the housing [Bro11]. The read/write heads on one side or on both sides of the disk touched this additional layer and worked principally like the first HDD heads: For writing, current flew through a coil in the head, magnetizing the magnetic particles in the respective area; for reading, the small voltage induced by the magnetic particles was amplified and detected. The formatting process which was necessary for each new disk aligned the formerly unordered magnetic particles to tracks with unused spaces between, to allow for small deviations of the rotation speed. Usual floppy disks had, e.g., capacities of 80 kB (the very first read-only 8 inch disks [IBM71]), 360 kB (5 ¼ inch DD), 1.2 MB (5 ¼ inch HD), or 1.44 MB (3.5 inch HD) [Bro11].

As a further development of Twistor memory, bubble memories were introduced in the early 1970s [Ros07]. In bands of orthoferrite, a material with easy magnetic axis oriented out-of-plane, small islands of magnetization can be built by a magnetic field perpendicular to the plane. These islands, the so-called bubbles, are perfectly circular – and they can be moved by an external magnetic in-plane-field, directed along defined paths by small ferromagnetic structures. In bubble memories, a steadily rotating field is formed by a coil network, driving the bubbles – and the "empty" places denoting the opposite bit value – along the material, similar to other memories with rotating disks or a rotating drum. Much smaller bubbles could be created in garnet which was thus used in bubble memories. With this material, the area necessary to cover ~ 4 kbits could be reduced from one square foot for ferrite-core memory to one square inch. However, the inventor, Andrew Bobeck, did not plan to compete against ferrite-core and other RAM, but configured the bubble memories as serial memories, similar to disk and drum memories. Nevertheless, bubble memories had some problems – with increasing bubble density, the influence of material impurities grew; skipping bad sectors was easier in hard disks; and finally, while hard disks had some time to develop further before the first semiconductor memories were invented, bubble memories lacked this development time.

A recent further development of bubble memory is the so-called Racetrack memory or domain-wall memory, being examined at IBM by a team led by Stuart Parkin [Par08]. For the Racetrack memory, an array of magnetic nanowires is arranged in horizontal or vertical orientation on a silicon chip. In such nanowires of about 100 nm thickness, domain walls can be moved by short spin-polarized current pulses, similar to bubble memories which used currents to drive magnetic patterns through a substrate. In this way, a "train" of ~ 10 to 100 domain walls can be read or written at the same time – thus, each nanowire stores several bits of data. If successful, the Racetrack memory could combine the high storage densities of hard disk drives with the high read/write speed of flash memory or the like.

While a 3 bit version could already be demonstrated [Hay08], there are still several challenges to be solved: In the first experiments, current pulses on the order of microseconds were necessary to move the domain walls, which is three orders of magnitude slower than expected. This effect could be attributed to imperfections in the nanowires, "catching" the domain walls [Mei07]. Without imperfections, the necessary pulse lengths were about some ns, as expected. Additionally, the voltage necessary to drive the domain walls is proportional to the wire length, restricting the possible amount of bits per wire.

Another candidate for a "universal memory" is the magnetoresistive random-access memory (MRAM) which has been developed since the middle 1990s [Ake05]. Opposite to conventional RAM which stores data in the form of electric charge (volatile), MRAM is a magnetic storage element (non-volatile). In the simplest form, an MRAM consists of a spin valve for each bit. Such a spin valve includes a permanent and a changeable ferromagnetic layer, separated by a thin insulator, thus building a TMR system [Khv12]. The resistance difference due to the relative orientation of both ferromagnetic layers defines whether a bit is set to 1 or to 0. Writing can be done by a grid of write lines, with those below the cells perpendicular to those above the spin valves – similar to the method used in core memories. An access transistor can be connected in series to the magnetic tunnel junction, blocking parasitic currents [Mül03]. Downscaling the single bits is limited by the induction fields erroneously reaching neighboring bits, leading to undesirable writing in wrong positions.

This problem could be overcome by using a spin transfer torque [Slo96, Ber96, Slo02, Sbi11, Kaw12, Thi12], i.e. spin-polarized electrons which would directly rotate the domains, thus reducing the currents necessary for writing [Vic12, Khv12], but at the same time creating the necessity of maintaining the spin coherence. This technique is developed further in most large semiconductor companies [Chu10, Yod10, Woo11, Gaj12] because it has the chance to combine the advantages of other techniques – the speed of SRAM (static random access memory), the density of DRAM (dynamic random access memory), unlimited endurance as well as low power consumption [Ake05]. Approaches for a reduction of the necessary power from recent values of about 10^6 A/cm² [Wan13] about two orders of magnitude can, e.g., be based on magneto-electric effects using electric fields / voltages to manipulate the magnetization (instead of the current) [Mar09, Wu11] by altering the interfacial perpendicular magnetic anisotropy of magnetic tunnel junctions [Wan12, Wan13] or by using the giant spin Hall effect for which pure spin currents are driven through magnetic tunnel junctions [Liu12].

Another approach is, similar to the Racetrack memory, the idea to store more than one bit per cell. This could be realized using a stack of magnetic tunnel junctions with different resistance levels, resulting in one bit per tunnel junction. For a cell with two bits, a two-step read/write process has been described [Kaw12]. In this process, firstly the resistance of the bit with smaller resistance is read using a mid level reference; the resistance state of the higher resistance cell is read with a higher or lower reference level, dependent on the result of reading the first bit. In writing, two bits with identical value can be written at the same time, while for two bits with opposite values, the bit with larger resistance has to be written first, opposite to the reading order.

For penta-layers including two pinned layers in antiparallel orientation and a free layer in the middle, separated by spacer layers on both sides, theoretical and experimental examinations have shown a decrease of the critical current density [Fuc05, Mak11] and the switching time [Mak11, Mak11a].

Other possibilities to reduce the necessary current and thus the minimum bit dimension are the thermally assisted switching which heats the desired area for a short time, or – similar to hard disk drives – a change to a perpendicular geometry [Ohn11, Sbi11a], a technique which still requires reduced damping and increased thermal stability [Mak12]. However, despite nearly 20 years of development, MRAMs have not yet proven to be advantageous to other memories.

Alternatives using no magnetic effects are, e.g, the FeRAM (ferroelectric random access memory) using the remanent polarization of a ferroelectric thin film, such as bismuth ferrite ($BiFeO_3$) and other perovskite type structures [Fuj06], and the PCRAM (phase change random access memory) in which two different phases of the storage material define the stored information bits [Mül03].

In magneto-optical disks, looking similar to floppy disks, the information is stored in a magnetic layer as in hard disk drives, and writing occurs thermo-assisted by heating a single spot with a laser and changing the local magnetization by an electromagnet, similar to the procedure described before. Reading, however, with a reduced laser power uses the magneto-optic Kerr effect [Bec98] which is described in chapter 3.1.

It should be mentioned that the different non-volatile RAM techniques have different properties and are thus more or less suitable for different applications: The write latency of spin-torque transfer memory, e.g., is about four times higher than in conventional DRAM, while the read latencies are comparable [Li12]. For PCRAM, the write latency is even times higher than in DRAM, the read latency still twice as long.

Additionally, the writing operation may require significantly higher energy per bit, e.g. for PCRAM the necessary writing energy is 50 times higher than in DRAM.

Finally, the writing endurance of, e.g., PCRAM is about a factor of 10^6-10^8 lower than the value for conventional DRAM [Li12].

Flash memory, however, is expected to reach scalability limits at the 20 nm technology nodes [ITRS]. Transistor-free cell structures, i.e. passive devices that switch by a change in the electrical conductance, are expected to be easier scalable [Ou11].

Amongst the mentioned technologies, only MRAM combines the speed of SRAM, the density of DRAM, and the lack of degradation, making it a candidate for a "universal memory" [Mun06].

These examples show that the new non-volatile techniques suffer often from different technical limitations, which suggest trying completely new ideas for future storage media.

2.3. Ideas for future magnetic storage media

While MRAMs and Racetrack memories could not yet enter the market in a substantial way, hard disk drives are still being developed further, to keep pace with the developing market of the SSD (solid state drives, using integrated circuits to store data). Similar to the idea described in the MRAM section above, heat-assisted magnetic recording can be used by utilizing a near field laser spot to punctually heat up the magnetic material, resulting in a decreased coercivity [TDK13]. Alternatively to a laser, MAMR (Microwave Assisted Magnetic Recording) uses a microwave pulse [Cou12, Zhu08].

With bit-patterned media [Ric06], Toshiba has announced the possibility to store about 2.5 Tbits per square inch, which would be about five times as much as in recent high-capacity hard disk drives [Tos10]. In such media, a self-assembled magnetic dot array is used instead of common thin film systems in which the poles of neighboring bits can be quite near, thus accidentally influencing each other.

A different possibility of using patterned media, including an architecture with x- and y-wires, are the nanowire-based resistive switching memories (NWRRAM) [Iel11, Iel13] or, more generally, nano-ionic memories [Asa97, Koz99, Bec00, Was07, Bor10]. In such an RRAM, the intersections of column wires and row wires represent the storage nodes, allowing for bit dimensions scalable to below 1 nm. The devices can be produced either top-down with optical lithography or bottom-up with self-assembling nanowires or nanodots [Hua01, Hua01a, Zho03]. Applying an electric field across the structure leads to ion diffusion and thus to the formation of a conductive path between the crossed wires which consist mostly of a metal core and a very thin oxide (i.e. non-conductive) shell [Li09, Cag11].

GMR (giant magneto resistance) heads using the cpp (current perpendicular to plane) geometry have already been tested for magnetic recording. In first experiments, they were found to allow for areal densities of about 400 Gbits/in² [Car08].

Shingled writing can be implemented easier at the device level [Ame10]. In this technique, writing tracks overlap, allowing for data in subsequent tracks to be destroyed during writing. This leads to possible smaller grain sizes. In a theoretical model [Gre09], a maximum data density of 3 Tbits/in² could be achieved, which is three times the superparamagnetic limit for conventional recording of 1 Tbit/in².

Recent research approaches showed the possibility to store one bit in only 12 atoms [IBM12]; however, for the commercial design of a hard disk using the bit-patterned media technique, the recent borders of lithography have to be taken into account. Thus, manufacturability of single features limits their dimensions to minimally 22 nm, a value which has already been reached [Kim10]. Recent research in lithography techniques aims at the 14 nm node [Kon12, Mar12, Ogi12].

Similar to the shingled writing described above or the bit-patterned media, new recording techniques are necessary to overcome the superparamagnetic limit ... or completely different ideas, such as the Racetrack memory.

A new idea, which is examined in this thesis, combines the ideas of bit-patterned media and Racetrack memory with more than one bit per magnetic feature. Similar to Racetrack memory, systems are searched which have the ability to store more than one bit per magnetic particle. While the Racetrack memory needs a completely new device, including new platters and new heads, it is tried to find a magnetic pattern which can be easily produced by nowadays lithography techniques, and which can principally be handled with available read/write heads. Opposite to MRAM and bubble memories, which could not really enter the market yet due to their complete incompatibility with most often used techniques, such as hard disk drives, changing from thin layers to bit-patterned media can be performed much easier. Thus, it is believed that magnetic particles with the capability of storing more than one bit, resulting from novel magnetic anisotropies, have a real chance to develop hard disk drives further, following Moore's law even over the restrictions of the superparamagnetic limit. A main idea of the approach presented in this dissertation thesis is derived from analysis of magnetic anisotropies and internal magnetic fields mainly tailored by sample shapes.

2.4. Technical interest in systems with novel anisotropies for data storage media

Some so-called exchange bias samples, consisting of a ferromagnetic and an antiferromagnetic layer, show not only a shift of the hysteresis loop when field cooled below the Néel temperature of the antiferromagnet, but also a modification of the loop shape. In the system Fe/MnF_2, e.g., a pronounced step can be seen on one side of the hysteresis curve [Ehr11a], as depicted in Fig. 2.3. Following this step back to numerically smaller external fields until the field vanishes gives rise to a stable intermediate state (green dots).

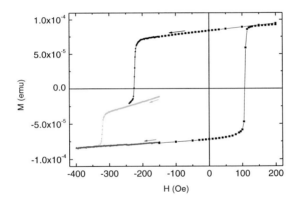

Fig. 2.3: Hysteresis loop, measured on an exchange bias sample made from Fe/MnF_2 at $T = 20$ K with a SQUID (Superconducting QUantum Interference Device). From [Ehr11a].

Such a stable intermediate state can be explained by a modified mathematical description for the anisotropy energy of the sample, as shown in Fig. 2.4: While the usual phenomenological description of the energy (cf. Appendix I) with a term proportional ($\sin^2\Delta\varphi\ \cos^2\Delta\varphi$) results in an energy landscape with smooth maxima and minima, the hard axes (at $45° + n \cdot 90°$) can be pronounced stronger by a mathematical description of the anisotropy energies proportional to $|\sin\Delta\varphi\ \cos\Delta\varphi|$ (Fig. 2.4, right panel). Although both descriptions lead to fourfold energy landscapes, there is an obvious difference in the resulting magnetization reversal processes – while for the usual phenomenological approach, magnetization reversal from $90°$ to $270°$

(chosen here for better visibility instead of the usual definitions 0° to 180°) is performed in one step (Fig. 2.4, left panel), the other mathematical description leads to a stable intermediate state, as can be seen in Fig. 2.4 (right panel).

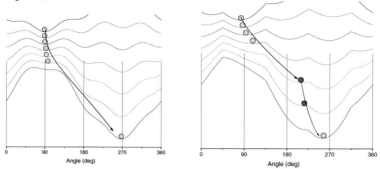

Fig. 2.4: Magnetization reversal for the usual phenomenological (left panel) and a possible different fourfold anisotropy (right panel), with the highest energies / largest positive external magnetic fields at the top and negative external fields at the bottom. From [Ehr11a].

The intermediate state and the correlated step in the hysteresis loop, however, have not been observed yet in thin film ferromagnetic samples, while exchange bias systems work normally only at low temperatures and only offer three possible states at vanishing external field (see Fig. 2.3). Thus, this work will examine nanostructured systems and possible anisotropies besides the usual phenomenological ones, to search for systems with four or even more stable states at vanishing external magnetic field.

3. Experimental results

In this chapter, the experimental setup for measurements of the Magneto-optical Kerr-effect is given, followed by a brief overview of other measurement techniques. The preparation of special textile-based samples and the experimental results of MOKE measurements on several samples in different crystalline orientations are described, alongside with comparisons of the experimental results with simulations using Magpar and MathCad. Measurements on nano-cylinders with asymmetric holes are described. Additionally, the textile-based samples with quite uneven surfaces have been tried to be measured by MOKE.

3.1. MOKE setup

The magnetization in a sample can be measured by a SQUID, as shown in Fig. 2.3. Such a SQUID can detect very small changes in the magnetization; however, the instrument is very large, expensive and works only with liquid helium which is an additional cost factor. Furthermore, measurements take several hours per hysteresis loop.

Another possibility to measure magnetization is the magneto-optic Kerr effect (MOKE). This method is less sensitive than SQUID measurements, but also significantly less expensive and much faster – MOKE measurements can be performed in less than a minute for a good setup and samples with high signals, and in correspondingly longer times for samples with very low signals.

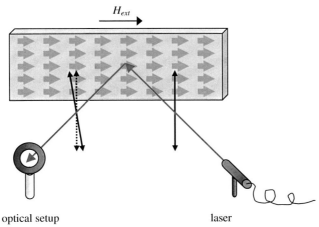

Fig. 3.1: Principle of MOKE measurement. For details see text.

The idea of MOKE measurements is depicted in Fig. 3.1: A linearly polarized laser beam impinges on a sample. The magnetization in the sample influences the rotation of the linear polarization axis in the reflected laser beam. If the magnetization in the sample changes, as it happens along a hysteresis loop, this procedure can be visualized by measuring the rotation angle of the reflected beam.

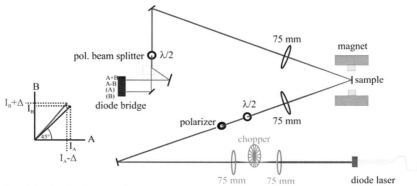

Fig. 3.2: Optical setup for MOKE measurements with sketch of diode bridge function (inset on the left).

Since this change in the polarization axis is very small (normally less than 1°), it has to be detected by a sophisticated optical setup, as depicted in Fig. 3.2.

Starting with the diode laser, a polarizer is used to enhance the polarization. The polarization direction can afterwards be rotated by a $\lambda/2$ retardation plate. This can be used to work with an exactly s-polarized beam, i.e. a beam with the polarization axis perpendicular to the plane of incidence/reflection. In this polarization direction, the longitudinal magnetization component (along the external field direction) can be unambiguously detected. On the other hand, rotating the polarization to an angle of 45° results in measuring the transverse magnetization component [Til06]. Afterwards, the laser beam is focused onto the sample which is placed inside a homogeneous magnetic field.

The next lens parallelizes the laser beam again before it is lead through a second $\lambda/2$ retardation plate and a polarizing beam splitter to the diode bridge. This bridge works in the following way: Starting, e.g., at positive saturation of a sample, the bridge signal "A-B" is set to zero by rotation of the $\lambda/2$ retardation plate until the ordinary and the extraordinary beam passing through the polarizing beam splitter have the same intensity. If the magnetization in the samples changes, a small positive or negative signal will be measured in the "A-B" channel (cf. inset in Fig. 3.2). This signal is significantly more exact than the usual measurement method in which the small intensity change of a large photo current in only one diode is measured.

Additionally, if the signals are very small, a chopper can be used (depicted in grey) which triggers a lock-in amplifier, to enhance the signal-to-noise-ratio.

3.2. Other methods to measure magnetization

Besides MOKE, there are several other methods to measure magnetization or magnetic anisotropies, respectively. In this sub-chapter, three other methods are described which have also been used within this thesis.

A SQUID (**s**uperconducting **qu**antum **i**nterference **d**evice) belongs to the most precise sensors of magnetic moment in a sample. It works due to the principle that the magnetic flux in quantized in a superconducting coil. Introducing a magnetic sample into the ring leads to an electric current in the ring which balances the flux to the next higher / lower magnetic fluxon. The voltage induced by an additional direct current through the ring depends on the compensation current, allowing for measurements of the magnetic field in the ring.

SQUIDs are mostly produced by sputtering or laser ablation [Cla04], often using classical superconductors such as Niob, which have to be cooled with liquid helium. SQUIDs based on high-temperature superconductors, which can work with liquid nitrogen, often have a higher noise and are normally more expensive.

BLS (**B**rillouin **l**ight **s**cattering) describes the inelastic scattering of a photon (laser beam) at a phonon, a magnon (magnetic spin wave) or other low-frequency quasi-particles, leading to a shift in the photon frequency which can be determined experimentally. This frequency shift is identical to the energy of the magnon (or phonon) created or destroyed in the scattering process. Due to the low interaction rate, experiments have to be carried out with a sophisticated optical setup, called Fabry-Pérot interferometer.

BLS can be used to measure magnetic anisotropies directly by rotating a given sample, since the magnon frequencies are correlated with the fields in the sample, including the angle-dependent anisotropy fields.

FMR (Ferromagnetic resonance) probes spin waves in ferromagnetic samples, giving rise to the magnetic anisotropies of a sample, as well as BLS does. The precession frequencies of the magnetization depend, amongst others, on the anisotropies in the sample. Due to changes in the linewidths and the possibility of additional frequencies occurring due to other effects, the evaluation of the raw data is much more complicated than in BLS.

Table 3.1: Comparison of different measurement techniques

Method	Advantage	Disadvantage
MOKE	Non-expensive, fast	Quite even surfaces necessary
BLS	Direct measure of anisotropies	Absolutely even surfaces necessary
SQUID	Exact measurements, uneven samples	Slow, expensive equipment
FMR	Direct measure of anisotropies	Analysis only by special software

3.3. Preparation of magnetic samples with textile methods

In a project supported by the Internal Project Funding of the Niederrhein University of Applied Sciences, magnetic samples have been prepared by typical textile production methods, such as warp knitting, winding, coating, or screen printing. Some of the samples have been fixed by lamination afterwards. Textile-based sample provided preliminary inspirations to patterned magnetic systems showing some static anisotropy properties at microscale characteristic for patterned memory devices.

Since textile-based samples cannot necessarily be expected to offer a smooth enough surface to be measured by MOKE or even D-MOKE (Diffracted Magneto-Optic Kerr Effect), a focused red laser ($\lambda = 632$ nm, $P = 1$ mW) has been used to create reflection patterns which have been photographed by a digital camera Dimage Z5 with super-macro. Such pictures can give information about possibilities of magneto-optical measurements.

Tab. 3.1 gives an overview about the different groups of experiments, which are described in the following.

Table 3.2: Textile-based magnetic samples (representative selection). For details see text.

Description	Microscopic view	Reflection pattern
Stainless steel wires diameter 28 µm in warp knitted fabric, polished on Struers LaboPol-5 with polishing spray 1 µm		
Stainless steel wires diameter 28 µm, wound around a flat substrate, coated only from the backside		
Transparency film coated with Ferricon 200 (Fe flakes with diameter 10-20 µm) by company Eckart, Hartenstein/ Germany		
"Reflectite" reflection tape without top layer, here shown before coating; half-sphere height ~ 20 µm, diameter ~ 50-60 µm		
Punched Ni rota-print plate for rotation screen printing, hole diameters 80-180 µm available		
Screen printing with flock glue, followed by strewing Fe particles, on several textile woven fabrics with 10-200 µm hole diameter		

- Samples containing fine stainless steel wires. The stainless steel wires in our experiments have diameter 28 μm and are made of austenitic steel with 10-13 % nickel (AISI 316 L, i.e. DIN 1.4404). Such steels are normally paramagnetic; this behavior changes due to the mechanic impact during pulling the fine wires to their final diameter, especially for low Ni contents. Tests with SQUID and FMR verify the ferromagnetic behavior of such wires and the clear uniaxial (two-fold) anisotropy in samples with parallel wires (Figs. 3.3, 3.4). For magneto-optic examinations, however, the surface is quite rough after polishing (Table 3.2, top panel). The relatively regular diffraction patterns of unlaminated and unpolished samples (Table 3.1, second row) indicate a chance to examine these samples by D-MOKE.
- Samples coated with magnetic material. For the coating, Fe flakes with diameter 10-20 μm "Ferricon 200", produced by the company Eckart, Hartenstein/ Germany, have been used. Additionally, Fe powder with particle diameter 10-20 μm has been used. To get a four-fold anisotropy, coatings with both forms of Fe pigments have been applied as thin films on woven polyester and cotton fabrics, additional to tests on smooth transparency films.
- As additional structured substrate with another anisotropy, the reflector tape "Reflectite" has been examined. With half-spheres of diameter ~ 50-60 μm and heights ~ 20 μm, this base material is similar to the non-magnetic half-spheres used in [Soa08, Ama10] as base for sputtering magnetic material on top. However, as can be seen from the microscopic image and the diffraction pattern of the uncoated Reflectite, the half-spheres are not regularly arranged.
- Rota-print nickel plate, with holes of diameter 80-180 μm in a regular six-fold pattern. Since this tool for a textile procedure (rotation screen printing) is magnetic itself, it has also been tested.
- Screen printing with flock glue on glass substrates through fine gauzes (hole diameters 10-200 μm) and gluing the gauzes themselves, followed by strewing Fe particles (diameter ~ 10 μm) on the glued parts. The idea of creating a four-fold pattern on the glass substrates or on the gauzes could in this way not be realized, since the glue tends to cover the complete substrate, not only the parts directly under open areas of the gauze. Only the Fe particles glued on gauzes of hole diameters 100-200 μm could be ordered in the four-fold pattern of the woven substrate, while even on finer gauzes, the glue built a closed film, resulting in a homogeneous Fe layer.

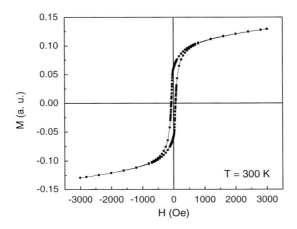

Fig. 3.3: SQUID measurement of a uniaxial sample produced from stainless steel wires with diameter 28 μm, wound around a flat substrate. A ferromagnetic hysteresis loop is clearly visible. From [Ehr11a].

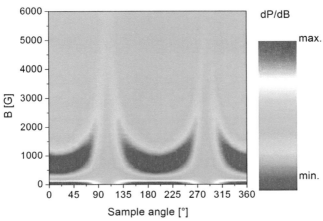

Fig. 3.4: FMR measurement of a uniaxial sample produced from stainless steel wires with diameter 28 μm, wound around a flat substrate. The uniaxial (two-fold) anisotropy can clearly be identified. From [Ehr11a].

Of the sample types described above, only two show a relatively regular diffraction pattern: the fine stainless steel wires wound around a flat substrate, and the rota-print plate. All other samples cannot be examined by D-MOKE, instead they have been tried to be measured by MOKE.

3.4. MOKE experiments on thin layer samples (Co/CoO)

The samples examined in this chapter have been produced during the author's diploma thesis by MBE (Molecular Beam Epitaxy) in the Phys. Inst. IIA, RWTH Aachen, Germany. They are constructed as follows:

- Substrate: MgO in the crystal orientations (111), (110), or (100)
- 1st layer: CoO 20 nm, crystal orientation identical with substrate
- 2nd layer: Co 6 nm, crystal orientation identical with substrate
- 3rd layer: cap layer to avoid oxidation

The 1st and 2nd layer have partly been exchanged, resulting in a "twinning" or other changes in the crystallographic structure of the CoO layer.

Since all measurements shown here have been performed at room temperature, i.e. above the Néel temperature of CoO of 291 K, the samples can be expected to behave similar to the pure ferromagnets. Thus, as first expectation, they should show a three-fold anisotropy on MgO(111), a four-fold anisotropy on MgO(100) and a four-fold plus uniaxial anisotropy for MgO(110) substrates.

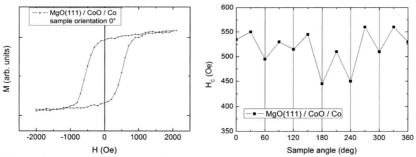

Fig. 3.5: Example hysteresis loop, measured on MgO(111) / CoO / Co for a sample orientation of 0° (left panel), and angle-dependent coercivities H_C for this sample (right panel).

Fig. 3.5 shows a typical MOKE measurement on a Co/CoO sample in (111) orientation at room temperature. Neither in this sample orientation nor in another angle, an exchange bias is visible, supporting the above mentioned idea that at this temperature only the pure ferromagnet is responsible for the shape of the hysteresis loop.

Since the Co layer is known to grow epitaxially in (111) orientation with fcc (face centered cubic) crystallographic type in this sample design [Til2001], a three-fold anisotropy might be expected first, as mentioned earlier. However, in this case some opposite orientation, e.g. 0° and 180°, would have to be different. This would imply that these hysteresis loops would have to be asymmetric, showing a horizontal shift similar to an exchange bias. This effect has not been observed. Thus, it can be assumed that Co(111) has a six-fold anisotropy. Fig. 3.5 (right panel) shows an angle dependence of the coercivity which fits to a six-fold anisotropy; however, more detailed measurements with a precise rotational sample holder have to support this finding in future examinations.

In Fig. 3.6, the results of a MOKE measurement on an MgO(110)/CoO/Co sample for an orientation of $0°$ is shown (left panel), with the typical shape – broad and rectangular-shaped – of an easy-axis loop. The right panel depicts the angle-dependent coercivities H_C. Here, the expected four-fold plus uniaxial anisotropy can clearly be seen.

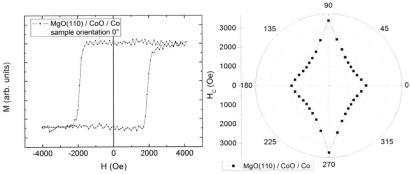

Fig. 3.6: Example hysteresis loop, measured on MgO(110) / CoO / Co for a sample orientation of $0°$ (left panel), and angle-dependent coercivities H_C for this sample (right panel).

In the same way, the sample MgO(100)/Co/CoO exhibits the expected four-fold anisotropy, as Fig. 3.7 shows.

Thus, the Co/CoO samples, measured at room temperature, give rise to a basic set of anisotropies which can be compared with the simulations depicted within this thesis. The anisotropies of other samples with different materials and geometries will be described in the further chapters.

In the next sub-chapter, a comparison of these experimental results with some simulations by Magpar can be found (finite element micromagnetics package invented at Vienna Technical University: http://www.magpar.net/). A more detailed introduction into the simulation software is given in Chapter 4.

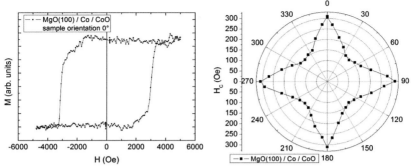

Fig. 3.7: Example hysteresis loop, measured on MgO(100) / Co / CoO for a sample orientation of 0° (left panel), and angle-dependent coercivities H_C for this sample (right panel).

3.5. Comparison with results of Magpar simulations

A comparison of the threefold Co/CoO(111) system and the simulation of the 3-wire system is given in Fig. 3.8. In both cases, the expected sixfold geometry cannot be detected clearly. While in the simulated values, small deviations from a perfect mesh geometry lead to significant changes between actually identical angles, the experimental deviations from a perfect sixfold angle-dependence can be attributed to imperfections in the sample, resulting in different coercivities for different spots on the sample.

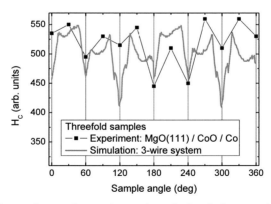

Fig. 3.8: Comparison of experimental and simulation results of threefold samples.

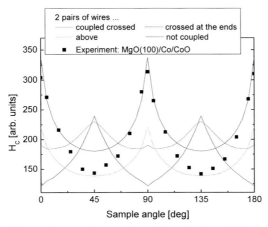

Fig. 3.9: Comparison of experimental and simulation results of fourfold samples.

Fourfold wire samples have been simulated in different geometries. Fig. 3.9 shows a comparison of the various samples with the experimental results of a measurement on an MgO(100)/Co/CoO sample which can also be expected to show a fourfold anisotropy.

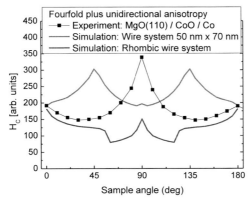

Fig. 3.10: Comparison of experimental and simulation results of fourfold plus uniaxial samples.

Interestingly, the non-coupled sample (cf. Fig. 4.2.5) has the most similar angular dependence as the experimental values. This finding can be interpreted in such a way that the Co/CoO at room temperature behaves like two perpendicular uniaxial systems. This idea will be examined more in detail in Chapter 5.

For comparison with the Co/CoO(110) system which includes a fourfold and a uniaxial (twofold) anisotropy, a rectangular wire system has been simulated in addition to the square systems depicted in Chapter 4.3.

As Fig. 3.10 shows, the idea of a rectangular wire system representing the rectangular (110) plane is *not* correct, since hard and easy axes are apparently exchanged. Instead, the simulation system diagonals should have the typical ration of $1:2^{1/2}$ (lilac plot in Fig. 3.10), leading, however, to a step which is not visible in the experiment. Different combination of fourfold and uniaxial anisotropies will also be examined in Chapter 5.

3.6. Comparison with results of MathCad simulations

Mathcad® is software for engineering calculations, produced by PTC® [PTC13]. Opposite to other software programs, such as Mathematica, the program mainly uses equations instead of programming code and can thus be used very well for the calculation of complicated or detailed equations without learning a specific programming language. Additionally, Mathcad is even able to handle physical units, if necessary.

In this work, Mathcad 13 has been used which is available in the Niederrhein University of Applied Sciences.

In Appendix 2, a detailed description of the Mathcad program written for this thesis is given.

In Chapters 3 and 4, samples with twofold, fourfold, sixfold anisotropies and with a combination of twofold and fourfold anisotropies (Co/CoO(110)) are examined. Here, these anisotropies are compared with the "typical" phenomenological description of the respective terms [Bla07], i.e.

$E_2 = K_2 \sin^2(\phi\_m)$ for the uniaxial (twofold) anisotropy;

$E_4 = K_4 \sin^2(\phi\_m) \cos^2(\phi\_m)$ for the fourfold anisotropy; and

$E_6 = K_6 \cos(6 \phi\_m)$ for the sixfold anisotropy.

To simplify comparison, the first easy axis for each graph in Fig. 3.11 is identical with 0°, the values of the anisotropy constants K_x differ partly.

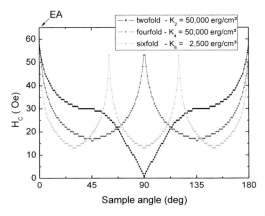

Fig. 3.11: Coercive fields, calculated for pure uniaxial (twofold), fourfold, and sixfold anisotropies.

The twofold (uniaxial) anisotropy results in a graph which is very similar to those produced by the micromagnetic simulation in Chapter 4; a more detailed comparison follows below. The graph found for the fourfold anisotropy is also comparable with those found in the experiment as well as in the micromagnetic simulations. The sixfold wire samples, however, led to slightly different angle dependences of the coercivity as the MathCad calculations. The triangular wire samples, finally, show a completely different angle dependence as the calculation in Fig. 3.11.

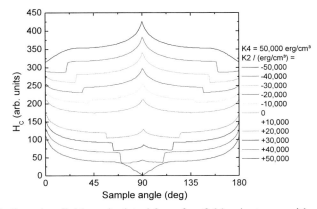

Fig. 3.12: Coercive fields, calculated for a fourfold anisotropy with additional uniaxial anisotropies with varying constants K_2.

For comparison with the Co/CoO(110) samples, a superposition of fourfold and uniaxial anisotropies has been calculated, the latter being varied. The results are depicted in Fig. 3.12. For large values of K_2, the uniaxial anisotropy dominates the angle dependence, while smaller values of K_2 result in mixed graphs in which a sharp step in the coercivity can be recognized.

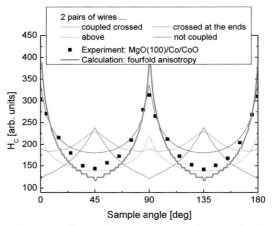

Fig. 3.13: Coercive fields for experimental examination and Magpar simulation of fourfold samples, compared with coercivity calculation of for a fourfold anisotropy.

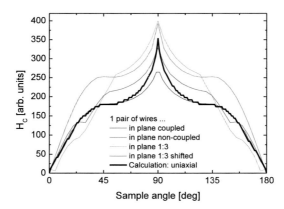

Fig. 3.14: Coercive fields for Magpar simulation of uniaxial samples, compared with coercivity calculation of for a twofold anisotropy.

As Fig. 3.13 shows, the MathCad calculation with a pure fourfold term of the form $E_4 = K_4 \sin^2(\phi\_m) \cos^2(\phi\_m)$ fits very well to the experimental results and quite well to the micromagnetic simulations of the systems without coupling between perpendicular wires (green and lilac lines). The coupled systems, however, show a different fourfold anisotropy. In Chapter 5.1, experiments with different anisotropy models are shown to match these forms better.

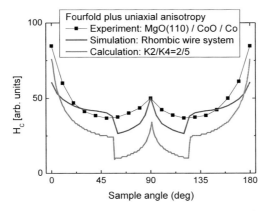

Fig. 3.15: Coercive fields for Magpar simulation of fourfold plus uniaxial samples, compared with coercivity calculation of a fourfold plus twofold anisotropy with a ratio $K_2 / K_4 = 2/5$.

For the combination of fourfold and uniaxial anisotropies, the calculation – here for a ratio of $K_2 / K_4 = 2/5$ – fits quite well to the simulation of the rhombic wire system (Fig. 3.15). The experimental results, however, are not reproduced by the first calculation with the phenomenological anisotropy model.

Concluding, the experimental results of the (110) samples as well as the micromagnetic simulations of four coupled wires are not modeled satisfactorily by the usual phenomenological descriptions of the respective anisotropies. Additionally, the simulation of the triangular wire system leads to completely different results as the sixfold anisotropy calculation, while some of the sixfold wire systems show at least similar results as the sixfold calculation. These deviations will be addressed in Chapter 5.1, where other mathematical descriptions of the respective anisotropies will be examined.

3.7. MOKE experiments on nano-cylinders

As a test for future MOKE measurements on systems of fourfold nano-structures (cf. Chapter 5.2), a sample consisting of nano-structured cylinders has been examined by MOKE. The sample, depicted in Fig. 3.16 (left panel), consists of arrays of 25-nm-thick permalloy dots with nominal diameters 1.0 µm arranged in a square lattice with a inter-dot-distances 1.0 µm, too. The circular void which is centered in one array and non-centered in the other one has a diameter of 160 nm [Vav06]. The sample has been prepared by the Vavassori group.

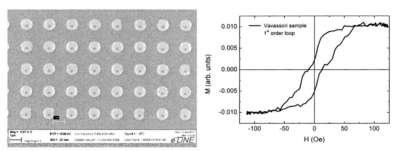

Fig. 3.16: Sample with nano-structured cylinders prepared by Vavassori *et al.* (left panel), 1st order hysteresis loop measured with D-MOKE (right panel).

Due to the lattice structure, not only usual MOKE measurements could be performed on this sample, but even D-MOKE measurements which measure not only in the direct reflection (0th diffraction order) but also in the 1st order, 2nd order diffracted beam etc.

Fig. 3.16 (right panel) shows such a 1st order result of a D-MOKE measurement. Apparently, the hysteresis loop is clearly visible and nearly as flat as in the original measurements by Vavassori *et al.* The other diffraction orders also gave results comparable with [Vav06] where experiments on the same sample are shown.

From these results, it can be concluded that our MOKE setup will allow for measurements on the future nano-structured samples to enable comparisons with the simulations depicted in this thesis.

3.8. MOKE experiments on textile-based samples with uneven surface

The samples depicted in Table 3.2 as well as several other textile-based samples have been examined by MOKE. While industrial steel samples have been shown to allow for MOKE measurements [Gon13], for most textile-based samples it was not possible to adjust the optical setup due to the low amount of reflected laser light. The planned way to overcome these problems by not focusing the laser beam on the sample (to measure an average of the rough and surely uneven surface) and collecting all light directly behind the sample did not result in a large enough amount of laser light on the photo diodes to be seen or even measured.

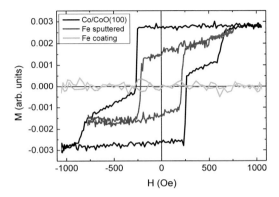

Fig. 3.17: Hysteresis loops, measured on an exchange-biased sample MgO(100) / Co / CoO (black line), an Fe sputtered thin film (red line), and a transparency film coated with Fe powder Ferricon 200 (cf. Table 3.2) (green line).

The only samples produced by a textile method, here coating, that allowed for a reasonable reflected laser beam were transparency films coated by the Fe powder Ferricon 200. Detecting a hysteresis loop, however, was not possible, as can be seen in Fig. 3.17. Comparison with typical loops of an Fe thin film and a Co/CoO epitaxial thin layer system shows that if there is a signal, it must be significantly lower than the other thin layer signals, while the noise is higher. If this measurement is possible, a more exact setup with lock-in amplifiers is necessary.

Since all textile samples and samples produced by textile-based methods did not give any signal in MOKE measurements, these experiments were stopped in favor of further simulations and efforts to get adequate nano-structured samples, which are described in Chapter 5.2.

4. Simulations with Magpar

In this chapter, the results of Magpar simulations are depicted. Starting with half-ball systems, their hysteresis loops, reversal dynamics and vortex core precessions, simulations have been carried on to systems consisting of one, two, three, four or six wires. This structure follows the route from classical bulk solutions, which are also used in recent efforts for bit-patterned media, to the new idea of wire systems (partly combined with half-balls at the ends) with different anisotropies.

In some of these systems, stable intermediate states have been found, which can be recognized by additional remanence states at zero external magnetic field. Correlated with the existence or absence of such intermediate states, different magnetization reversal processes have been identified. The influence of the wire lengths and diameters on the magnetic properties have been examined for four-fold wire systems, while in the six-fold systems, the corner shape has proven to strongly influence the magnetization states and reversal processes.

4.1. MagPar – a short overview of the program

Magnetic nanoparticles are on the one hand too small to ignore quantum-mechanical effects and on the other hand too large to be treated by ab-initio computational models. Micromagnetic models fill the gap between macroscopic and atomic models and allow for detailed calculations of not only the experimentally accessible values, such as a complete hysteresis curve, but also for the simulation of the time-resolved spatial distribution of the magnetization. Making use of a finite element method (FEM), all geometries can be modeled.

Magpar is a freely available micromagnetic simulation [Sch03] which has been used for the simulations in this chapter. It can handle different uniaxial and cubic anisotropies, exchange interaction, magneto-elastic effects and static as well as dynamic external magnetic fields. It is based on the dynamic integration of the Landau-Lifshitz-Gilbert (LLG) equation of motion [mag13].

The output files include a log file containing the time-dependent field sweeping sequence, the calculated values of the three magnetization directions, total and external energy as well as demagnetization and anisotropy energy. Png files show "snapshots" of the three magnetization fractions. Gz file additionally include the complete three-dimensional magnetization information for each node of the meshed sample.

Together with the inp file containing the mesh coordinates, the gz files can be used to create three-dimensional depictions of the magnetization in the sample at a certain time during the magnetization reversal process. These pictures often provide an inside into the reversal mechanisms, which could only be guessed by the pure hysteresis loops. Thus, they will be shown for nearly all magnetization reversal processes in this thesis. The pictures are created by ParaView, an open-source, multi-platform data analysis and visualization application which has been developed especially for very large data sets [Par13].

Preprocessing, i.e. the meshing of the desired nano-structures, is performed by GiD, the "personal pre- and postprocessor", developed by the International Center for Numerical Methods in Engineering (CIMNE) [GiD13]. Here, the node distances have to be taken into account according to the exchange lengths of the simulated materials, to model the physical properties of the samples properly.

4.2. Low-dimensional half-ball systems with shape modifications

Several magnetic states, typical for low-dimensional objects, like vortex, onion, or horseshoe states, have been examined extensively in recent experiments [Cow99, Zhu00, Zhu04, Zha10, He10]. Disks with different shapes have been studied in experiment and theory in order to test methods of magnetization dynamics control [Rot01, Klä03, Vaz05, Bed07, Elt10, Kim10a]. The analysis of flux-closed vortex states (i.e. vortices without core) is of special importance for applications of magnetic nano-particles in storage devices to reduce stray fields [Zha10, Zhu04]. Former research has proven that the formation of a vortex state strongly depends on dimension and shape of magnetic nano-particles [Cow99, Zha10, He10].

Nanostructures ferromagnetic (FM) objects have been studied recently in several theoretical and experimental studies [Bad06, Li10, Kod99], due to their importance in data storage and other applications [Ros01, Ter05, Ake05]. Their magnetic properties depend strongly on the nanoparticles' shapes and dimensions [Nog05], which enables tailoring the desired properties by changing the nanoparticle form. Recent efforts often concentrate on 2D nano-devices, such as dots [Red10, Hie97], rectangles and triangles [The10], rings [Zha10], wires [Cow02], tubes [Uso07], and mixed geometries [Hua10]. However, first experiments using thin magnetic films on non-magnetic half-spheres with diameters between 20 nm and 1,000 nm have shown the importance of probing 3D nano-magnets [Soa08, Ama10]. Such 3D nano-objects can be structures with recent self-assembling techniques [Leo10].

For small magnetic elements, shape modifications play an important role in magnetization dynamics; however, due to the reduced dimensionality, demagnetizing fields can intrinsically compete with exchange fields resulting in specific magnetic behavior – e.g. oscillations with one or more frequencies, rapid transient states, and static magnetic states at remanence or saturation.

For the simulations in this chapter, ferromagnetic permalloy (Py) half-balls with different shape modifications have been chosen. The circular base with diameter 100 nm is located in the x-y-plane. A perpendicular hole of diameter 50 nm has been cut in some of the half-balls to extract the core region of a possible vortex state which has a comparable diameter [Cho07]. The magneto-crystalline anisotropy has been excluded in order to concentrate on the competition between magneto-static and exchange-based interaction, although in nano-sized polycrystalline elements, the crystallite borders can induce additional local magneto-crystalline anisotropies.

Simulations have been performed based on the LLG equation of motion using MagPar simulator [Sch03]. The meshing uses tetrahedral elements not larger than 3.7 nm, which is smaller than the Py exchange length of 5.7 nm [Smi89], with an approximately ten times denser mesh along the edges to include the influence of demagnetizing fields more exactly.

The half-ball systems have been examined with the external magnetic field along the high symmetry direction (z-axis) or in the x-y-plane. The field sweeping speed of 10 kA/(m ns) is comparable to typical values in magneto-electronic application [Teh00]. Since different values for the Gilbert damping constant α of permalloy can be found in literature, two different values have been chosen for different simulations; a comparison is given in Chapter 4.2.d, while the influence of the field sweeping speed is examined in 4.2.e.

Parts of the results presented in this chapter have been reported in [Bla10] and [Bla12].

Each sub-chapter starts with an overview about the relevant parameters of the respective simulations, with the changing parameters marked by bold letters.

4.2.a. Hysteresis loops
The following parameters have been chosen for simulations:
Dimensions: Diameter 100 nm, height 50 nm, shape modifications
Material: Permalloy (Py)
Exchange constant A: 1.05 x 10^{-11} J/m
Magn. polarization at saturation J: 1 T
Gilbert damping constant α: 0.01

Tetrahedal mesh dimension: ≤ 3.7 nm
Field range: - 600 kA/m … + 600 kA/m (field along z-axis)
          - 450 kA/m … + 450 kA/m (field along x-axis)
Field sweeping speed: 10 kA/(m ns)

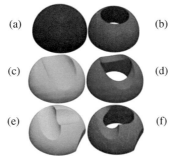

(a) (b)
(c) (d)
(e) (f)

Fig. 4.2.1: Simulated samples: (a), (c) and (e) are solid half-balls, while (c) and (d) half-balls possess a 50 nm diameter hole. The height of the half-ball (a) equals 50 nm. The cylindrical cuts in the x-y plane (b, d) have diameter 50 nm, and the maximum depth of a cut, seen along the half-ball z-axis, equals 25 nm with respect to the top point of the bulk half-ball (a). Samples (e) and (f) possess arbitrary, nonsymmetrical cuts, and the elliptical hole in (f) case is not parallel to the z-axis.

Fig. 4.2.1 shows the simulated samples. A detailed description is given in the figure caption. While the vortex core is excluded in samples b), d) and f), the samples e) and f) show the strongest symmetry breaks.

In Fig. 4.2.2, the hysteresis loops obtained for the external field Hext applied along the x-axis are shown. For the solid half-ball (a), it should be noticed that the magnetization reversal starts before $H_{ext} = 0$ is reached. Such a behavior has been described, e.g., in [He10] and can be interpreted as beginning of the vortex state formation. Snapshots of the magnetization in z-direction verify the correlation of the vortex formation with the beginning of the oscillations. A detailed analysis is given in Chapter 4.2.c.

Cutting the top of the solid half-ball suppresses the magnetization reversal before $H_{ext} = 0$ (c), while a hole in the solid half-sphere (b) significantly reduces the oscillations – in the area of the step (~ -20 kA/m … -180 kA/m), the magnetization builds a vortex state, like in the same field region in the solid half-ball, but since the vortex core region is excluded, no vortex core precession is visible. Opposite to this flux-closed vortex state, both saturation magnetizations and the oscillatory region are associated with onion states.

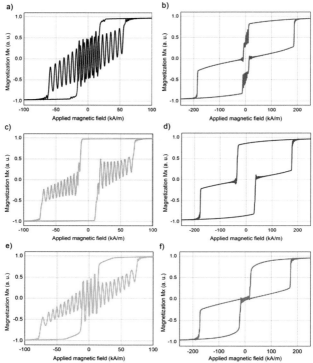

Fig. 4.2.2: Hysteresis loops, simulated for the samples depicted in Fig. 4.2.1, for the external field applied along the x-axis. From [Bla12], modified.

The snapshots of the magnetization reversal from positive to negative saturation shows the z-components of the magnetization, indicating a symmetric oscillation with the two colored domain walls oscillating towards each other and back again. Following the snapshots of the x-component of the magnetization from negative to positive saturation, the magnetization reversal from the saturated onion state along a second onion state to a vortex state and finally to a reversely saturated onion state is visible.

Additional top cuts in this sample with a hole reduce the coercivity again (d), opposite to the finding in solid half-balls. Asymmetric top cuts, finally, in the solid half-ball (e) and the half-ball with a hole (f) reduce the oscillations in the solid half-sphere, as well as the symmetric cut (c), and can even introduce the beginning of the vortex state before $H_{ext} = 0$ is reached in the sample with a hole (f). All these effects have to be taken into account if a certain oscillatory behavior, coercive field or remanence is desired.

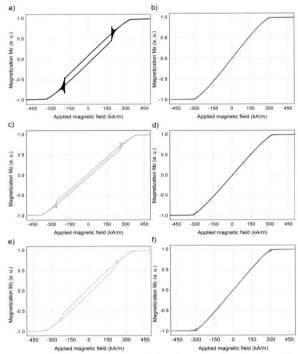

Fig. 4.2.3: Hysteresis loops, simulated for the samples depicted in Fig. 4.2.1, for the external field applied along the z-axis. From [Bla12], modified.

The results of simulations with H_{ext} along the z-axis are presented in Fig. 4.2.3. In the magnetization curves of the solid half-spheres without hole (a, c, and e), "open" hysteresis loops are visible, with weak oscillations around the magnetization reversal region; while for the half-balls with a hole (b, d, and f) completely smooth loops can be recognized which are apparently reversible in all field regions, like in superparamagnetic materials. Such reversible magnetization curves of quasi-spherical Fe particles of diameter 200 nm have also been described in [Dia10].

This behavior can be explained as follows: In the solid half-balls (without or with cuts), a vortex-core is built, which is separated from the outer part of the vortex by a circular domain wall. During the magnetization reversal process, the core region decreases, while the outer ring reverses. The oscillatory step finalizes the reversal by switching the core region. The cylindrical cuts in the solid half-balls (c, e) lead to smaller coercivities and remanences. Since the inner part, where the vortex core occurs, is missing in the samples with holes, these particles only show a smooth magnetization reversal in the ring. Thus, only complete half-balls without a central hole could be utilized in storage devices if the external magnetic field is to be applied along the z-direction, since only in these samples two different remanence states are available. Comparing Fig. 4.2.2 and Fig. 4.2.3, it is obvious that for the Py half-balls the hard axis is always directed along the z-axis, as in 2D nano-magnets.

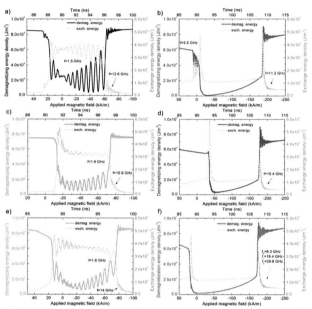

Fig. 4.2.4: Demagnetizing and exchange energy, simulated for the samples depicted in Fig. 4.2.1, for the external field swept along the x-axis from positive to negative saturation. From [Bla12], modified.

Figs. 4.2.4 and 4.2.5 depict the time-resolved demagnetizing and exchange energy for decreasing external magnetic fields, equivalent to the hysteretic evolution from positive to negative saturation (cf. Figs. 4.2.2 and 4.2.3).

In the saturated states, the demagnetization energy is maximal, while the exchange energy is minimal. Between these extremal states, different scenarios can describe the evolution of both energy fractions. In case of flux-closed states (Fig. 4.2.6), normally the exchange energy dominates. Demagnetizing fields, however, can significantly change their spatial distribution due to shape modifications, influencing the exchange fields in a similar way (Fig. 4.2.7). This affects the whole sample, the volume part as well as the edge regions, due to the reduced dimensions.

Fig. 4.2.4 illustrates the energy evolution for the external magnetic field along the x-axis. For the samples without a hole (a, c, and e), the oscillations of the vortex core are visible after passing the remanence. The frequency of 1.5 GHz for a perfect solid sample (a) is lower than 1.9 GHz for the sample with a horizontal cut (c) and also slightly lower than 1.6 GHz for the imperfect sample (e). The exchange and demagnetizing oscillations are anti-phased; however, the amplitudes of these oscillations are relatively weaker for the shape-modified samples (c) and (e). Importantly, the switching time is shorter for the sample with a cut (Fig. 4.2.4c) – where it is approximately 6 ns – than for the perfect solid sample (Fig. 4.2.4a), where it equals about 9.5 ns. This finding underlines the importance of imperfections for the utilization of such nano-systems in magnetic storage media, in which short switching times are important.

In the samples with a hole (b, d, and f) excluding the vortex core, the dynamics are different. The beginning of the intermediate-region evolution can happen via 3.5 GHz oscillations (b) or, for samples with cuts (d, f) without oscillations. These pre-intermediate oscillations are caused by the formation of flux-closed states induced at the circumference edges, similar to the behavior of the solid half-ball depicted in (a); however, they are eliminated by magnetic poles located at intentionally introduced edges, also for samples without holes (c, e). While there are no energy oscillations in the intermediate regions for the samples with holes, a rapid transient state exists for the exchange energy which can be attributed to the energy necessary to switch from the second onion state to the vortex state. The highest peak is observed for the sample with the horizontal cut (7.1×10^4 J/m^3). Importantly, this peak is reduced for the imperfect sample (f), where the randomly oriented magnetic poles, located at the edges, act against this effect.

The intermediate regions described above are followed by relatively high-frequency oscillations in the range of 10-30 GHz. The amplitudes of these oscillations are lower than those of the narrow-frequency (below 2 GHz) intermediate-region oscillations. Generally, it seems that modifications of the shape cause an increase of these ripple frequencies, sometimes even leading to the occurrence of higher harmonics (Fig. 4.2.4f).

In Fig. 4.2.5, the time-dependent energies are shown for the external magnetic field along the z-axis. Since the magnetic field direction is parallel to the high-symmetry axis, regular, closed circulations of the magnetization can be observed (see Fig. 4.2.6).

For some of the samples with holes (b, d), the symmetry and the smooth slope of the energies correspond to the completely reversible magnetic hysteresis loop. Only for the asymmetrically deformed sample (f), 2.4 GHz oscillations occur, followed by saturation. This behavior is associated with a competition between the demagnetizing fields, located near the external circumference edge, and the vortexes leaving the sample volume near that edge. In the three reversible cases for the nano-dots with holes (b, d, f), both energies show roughly opposing values – this effect may be attributed to the lack of energy loss in a hysteretic loop.

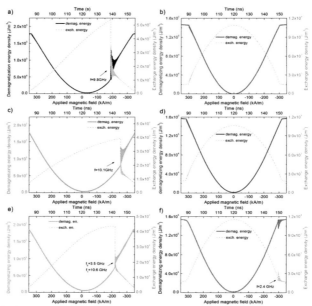

Fig. 4.2.5: Demagnetizing and exchange energy, simulated for the samples depicted in Fig. 4.2.1, for the external field swept along the z-axis from positive to negative saturation. From [Bla12], modified.

For the solid samples without hole (a, c, e), asymmetrically located oscillations can possess composed nature. Thus, the exchange energy maximum is phase-shifted in comparison to the demagnetizing energy minimum. The following ripple oscillations return the correlated anti-phased behavior, before the samples are saturated. This effect is shape-dependent, and the observed frequencies are in the range of 9.8-10 GHz. The reversal process is associated with the competition between demagnetizing poles, located not only at sample circumference, but additionally at newly introduced edges. As expected, the amplitude of these oscillations is smaller for the more deformed sample (e) in comparison to the more symmetrical cases (a, c).

Fig. 4.2.6: 3-dimensional visualization of magnetization vectors for the external magnetic field applied along the x-axis, for the solid half-ball (cf. Fig. 4.2.1a). Colors represent local values of M_x magnetization components; red: parallel to x-axis, blue: antiparallel to x-axis.

Fig. 4.2.6 shows the magnetization vectors from positive to negative saturation of the solid half-ball (a) for an external magnetic field along the x-axis. Starting from positive saturation in the left pair of sketches, the next pair depicts the beginning of the vortex formation, while the 3rd and 4th pair show the precessing vortex, before negative saturation is reached in the last pair.

Fig. 4.2.7 depicts the magnetization vectors for the external magnetic field along the x-axis of the solid half-ball with a hole. After the positive saturated state (left pair), a horseshoe-state appears, followed by the flux-closed vortex state which leads to negative saturation via a horseshoe-like state with strong out-of-plane magnetization components (4th pair).

Fig. 4.2.7: 3-dimensional visualization of magnetization vectors for the external magnetic field applied along the x-axis, for the solid half-ball with a cylindrical hole (cf. Fig. 4.2.1b). Colors represent local values of M_x magnetization components; red: parallel to x-axis, blue: antiparallel to x-axis.

Fig. 4.2.8. 3-dimensional visualization of magnetization vectors for the external magnetic field applied along the x-axis, for the solid half-ball with a top cut (cf. Fig. 4.2.1c). Colors represent local values of M_x magnetization components; red: parallel to x-axis, blue: antiparallel to x-axis.

In Fig. 4.2.8, the magnetization for sample (c), a half-ball with a top-cut, is depicted. While at first glance, the vortex formation is similar to the sample without imperfections, the vortex core is slightly trapped here in one half of the sample, with a barrier defined by the cut in which the lower sample height blocks the out-of-plane magnetization of the vortex core.

The simulation results described above give an overview about the magnetization characteristics which can be tailored by changes in the geometry of magnetic nano-particles, enabling a deeper understanding of the desired properties in a relatively broad range of modifications to meet the challenges of new 3D technology of magnetic devices. This could give an opportunity for new applications in magneto-electronics.

The next chapter will concentrate on the reversal dynamics in similar half-balls, introducing new shape modifications, allowing for an overview about the range of possible oscillation frequencies.

4.2.b. Reversal dynamics

The following parameters have been chosen for simulations:

Dimensions: Diameter 100 nm, height 50 nm, shape modifications

Material: Permalloy (Py)

Exchange constant A: 1.05 x 10^{-11} J/m

Magn. polarization at saturation J: 1 T

Gilbert damping constant α: 0.1

Tetrahedal mesh dimension: ≤ 3.7 nm

Field range: - 600 kA/m … + 600 kA/m (field along z-axis)

            - 450 kA/m … + 450 kA/m (field along x-axis)

Field sweeping speed: 10 kA/(m ns)

Opposite to the previous chapter, the Gilbert damping constant α is here set to 0.1. This stronger damping reduces the oscillations which dominated the results of the simulations on half-balls without hole for the external field along the x-axis, when α = 0.01.

The simulated Py half-balls are modified in the following way (Fig. 4.2.9): While sample (a) is an unmodified solid half-sphere again and sample (b) contains a symmetric hole of diameter 50 nm, the samples (c) and (d) have a roughened surface, and the former edges in the x-y-plane are cut.

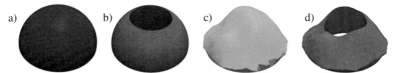

Fig. 4.2.9: Simulated Py half-balls: a) regular solid, b) regular half-ball with a hole, c) disturbed solid, and d) disturbed half-ball with a hole. The cylindrical hole diameter equals 50 nm. The diameter of the undisturbed half-ball base in the x-y plane is 100 nm, while the half-ball height, measured along the z-axis, equals 50 nm. From [Bla10], modified.

The time-resolved developments of the magnetization are depicted in Figs. 4.2.10 and 4.2.11. In Fig. 4.2.10, a field sweep along the z-axis is examined, with the z-component of the magnetization illustrated by the black line (left ordinate) and the x-component given by the red line (right ordinate). The insets show snapshots of M_z at several moments. The "central" time 60 ns (Fig. 4.2.10) or 50 ns (Fig. 4.2.11), respectively, is identical with the external field $H_{ext} = 0$.

Despite the different shapes, leading to different magnetization reversal processes, Fig. 4.2.10 shows some repeatable effects. Firstly, each "jump" in the magnetization along the z-direction is accompanied by an oscillation in M_x. However, there is no oscillation visible in M_z. The absence of oscillations in M_z is similar to the results of the former chapter, where for the external field along the z-axis only small oscillations were visible. It can be assumed that the complete suppression here can be attributed to the enhanced damping constant.

Secondly, the oscillation frequencies of M_x follow roughly the effective field intensity, with the lowest frequencies near $H_{ext} = 0$ (60 ns). A more detailed look on the observed GHz oscillations shows that they can be subdivided into two types: those close to saturation, and those correlated with the switching moment (between saturations), when a magnetization component changes its sign from positive to negative values, sometimes revealing a rapid, transient character. Additionally, it should be mentioned that oscillations can possess one or two dominating frequencies (see, e.g., sample (b) between 90 ns and 120 ns).

While for thin magnetic dots with an aspect ratio of ~ 10 or more, frequencies of some 100 MHz are reached [Gus02, Nov05, Gus06], the half-balls examined here (with a third dimension in the same order of magnitude as the lateral dimension) show significantly higher precession frequencies in the order of magnitude of some GHz. This is correlated to the possibility of a significantly faster vortex core reversal than the order of a few ps found in flat dots [Wae06, Her07], i.e. faster switching of the information in a dot representing a bit. Future simulations will examine the possibilities of ultrafast magnetic field pulses to trigger the vortex core reversal in these systems.

As has already been recognized in the previous chapter, intentional perturbations of the sample shape, influencing the demagnetization fields, may strongly suppress the magnetization oscillations. The holes, finally, seem to significantly influence the character (amplitude, duration and frequencies) of the oscillations. Thus, the perturbation of the sample shape can change oscillations resulting from demagnetizing fields, similarly to holes excluding vortex oscillations.

As has already been recognized in the previous sub-chapter, intentional perturbations of the sample shape, influencing the demagnetization fields, may strongly suppress the magnetization oscillations. The holes, finally, seem to significantly influence the character (amplitude, duration and frequencies) of the oscillations. Thus, the perturbation of the sample shape can change oscillations resulting from demagnetizing fields, similarly to holes excluding vortex oscillations.

To complete the above analysis, it is worth mentioning transient and rapid changes of magnetization, for example, that of M_x reaching an approximate value of 0.021 for $t = 82$ ns, shown in Fig. 4.2.10a. This situation is associated with maximized exchange energy of the system resulting from the existence of a vortex state. For $t > 82$ ns the vortex core begins to vanish, while in the solid half-ball with distortion (Fig. 4.2.10c), the vortex core starts moving along a direction favored by the uneven shape. This is why the effect is significantly reduced for the cases with 50 nm holes where a flux-closed vortex appears which is favorable for applications in magnetic storage devices.

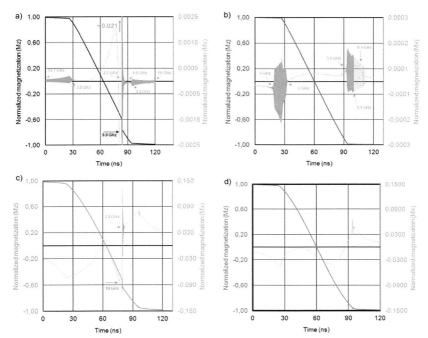

Fig. 4.2.10: Time-dependence of magnetization components M_z (left ordinate) and M_x (right ordinate) for the magnetic field applied along the **z-axis**, corresponding to the simulated shapes depicted in Fig. 4.2.9. Additionally, some characteristic frequencies are shown. The external field changes sign at 60 ns. From [Bla10], modified.

Similar findings can be observed for the external magnetic field along the x-axis (Fig. 4.2.11). Here, in each sample a two-stage magnetization reversal with oscillations in both magnetization components can be observed. These oscillations, however, are strongly reduced in comparison with $\alpha = 0.01$.

The holes lead to a strong reduction of the oscillations again, the enhanced surface roughness results in changes of the coercivity for the samples with and without hole. Sample (d) combines both advantages by exclusion of a vortex core and perturbation of demagnetizing fields, resulting in the strongest suppression of oscillations.

Similar transient effects as for the external field along the z-axis are observed here, exhibiting the nucleation and propagation of domain walls. Additional spikes are visible for all four samples. In these cases the transient effects are associated with maximized demagnetization energy which accumulates at the edges independently of the sample shapes. The results obtained for switching times and oscillatory behavior suggest that moderate not-perfectness of nano-elements can be an advantage for utilization in magnetic storage elements. All the above can give practical hints for designers of switching magneto-electronic elements.

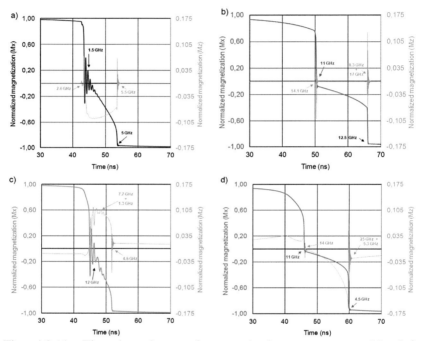

Fig. 4.2.11: Time-dependence of magnetization components M_x (left ordinate) and M_z (right ordinate) for the magnetic field applied along the **x-axis**, corresponding to the simulated shapes depicted in Fig. 4.2.9. Additionally, some characteristic frequencies are shown. The external field changes sign at 50 ns. From [Bla10], modified.

4.2.c. Vortex core precession and M rotation

Dimensions: Diameter 100 nm, height 50 nm, shape modifications
Material: Permalloy (Py)
Exchange constant A: 1.05 x 10^{-11} J/m
Magn. polarization at saturation J: 1 T
Gilbert damping constant α: 0.01

Tetrahedal mesh dimension: ≤ 3.7 nm

Field range: - 600 kA/m … + 600 kA/m (field along z-axis)

           - 450 kA/m … + 450 kA/m (field along x-axis)

Field sweeping speed: 10 kA/(m ns)

For a closer look on the vortex core precession during the magnetization reversal for an external field along the x-axis, the smaller damping constant α = 0.01 has been chosen again, as in Chapter 4.2.a. A comparison with the results for α = 0.1 will be given in the next chapter.

The vortex core precession has been examined by evaluation of the graphical snapshots of the z-component of the magnetization, produced by Magpar. For this purpose, the single snapshots have been copied into a coordinate system in Origin, and the position with the maximum intensity of red color has been noted. This procedure has been carried out three times to increase the evaluation accuracy.

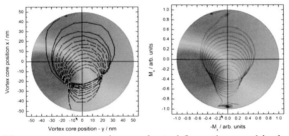

Fig. 4.2.12: Vortex core precession evaluated from the graphical snapshots of the simulation (left panel); M_x-M_y-plot of the simulated magnetization (right panel). In both graphs, the x-axis is directed to the top, since the snapshots (see colored disc in the background) are oriented in this way. The light-blue points correspond to snapshots of simulations for $H_{ext} < 0$.

Firstly, Fig. 4.2.12 (left panel) shows that the vortex core does not build and vanish in the middle of the half-ball, as it was visible, e.g., in Fig. 4.2.8 for the external field directed along the z-axis. Instead, here the core starts next to the edge which is nearest to the positive x-axis, which can be explained by the positive field which is still present at the beginning of the vortex state phase. Equivalently, the core vanishes around the edge near the negative x-direction, i.e. the direction of the external magnetic field at this time (light-blue points depict simulations in the negative magnetic field range).

Comparing the vortex core precession with the M_x-M_y-plot of the simulated magnetization (right panel), both graphs show very similar curves. A closer look, however, shows also differences: The saturation magnetization (i.e. M_x = ± 1) is naturally not reached within the vortex state phase, but can clearly be differentiated from the precession area. Additionally, in the M_x-M_y-plot of the vortex state, the magnetization precession phase takes less space than the vortex core precession. This can be easily explained by the snapshots in the background of Fig. 4.2.12: In the vortex core phase, with the vortex marked red here, equivalent to the magnetization directed along the positive z-direction, there is always a "dip" with opposite magnetization (blue area, depicting the magnetization oriented in negative z-direction) which is "dragged" behind the core [Kas06]. Thus, the overall magnetization, depicted in the M_x-M_y-plot, must be reduced in comparison to the vortex core position which is not influenced by the "counterpart" magnetization fraction.

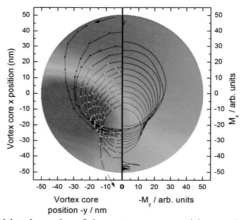

Fig. 4.2.13: Combined graphs of the vortex core position and the M_x-M_y-plot.

In Fig. 4.2.13, both plots of Fig. 4.2.12 are combined to allow for an easier comparison of both values. Evidently, the M_x-M_y-plot encloses a smaller "area" than the vortex core position plot.

The same effect is visible in Fig. 4.2.14, where the x- and y-values of the magnetization and of the normalized vortex core position are depicted vs. the external magnetic field. Here it can be clearly seen that although the amplitudes of both values differ, the frequencies are identical. This underlines that the oscillation which is visible in the hysteresis loops (Fig. 4.2.2) is strongly dominated by the vortex core precession; however, the magnetization in the residual part of the simulated samples can also have a fraction in z-direction, leading to a slight mismatch of vortex core position and M_x-M_y-plot.

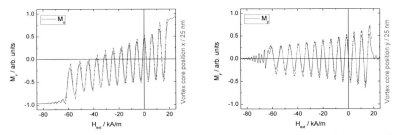

Fig. 4.2.14: Comparison of x- (left panel) and y-values (right panel) of magnetization (black, left ordinate) and relative vortex core position (red, right ordinate).

4.2.d. Influence of the Gilbert damping constant α

Dimensions: Diameter 100 nm, height 50 nm, shape modifications
Material: Permalloy (Py)
Exchange constant A: 1.05 x 10^{-11} J/m
Magn. polarization at saturation J: 1 T
Gilbert damping constant α: 0.01 / 0.1
Tetrahedal mesh dimension: ≤ 3.7 nm
Field range: - 600 kA/m … + 600 kA/m (field along z-axis)
              - 450 kA/m … + 450 kA/m (field along x-axis)
Field sweeping speed: 10 kA/(m ns)

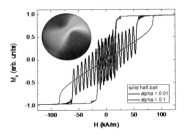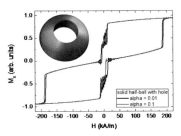

Fig. 4.2.15: Comparison of simulations with α = 0.01 (black lines) and α = 0.1 (red lines) for a solid half-ball (left panel) and a half-ball with cylindrical hole (right panel).

In Chapters 4.2.a and 4.2.b, the strong influence of the Gilbert damping constant α on the results of a simulation became already evident. The impact of the choice of α on the simulated hysteresis loops is examined here in detail for a solid half-ball and a sample with a cylindrical hole in the center of the half-ball, but without further shape modifications.

In Fig. 4.2.15, the effect of the choice of α is clearly visible. For the solid half-ball, the oscillations are significantly suppressed by the larger value of α. Interestingly, the smooth parts for α = 0.1 in the field regions between ~ 20 kA/m and ~ 90 kA/m and the same region with negative sign are approximately identical with the average magnetization in the oscillatory part for α = 0.01. For this special sample and the simulation conditions given above, however, the oscillatory behavior for α = 0.01 leads to a faster magnetization reversal from the vortex state into saturation. Apparently, the strong precessions provide a possibility to overcome the energy barrier between those states, similar to an external field pulse which can induce the so-called "precessional switching". Nevertheless, the first step in the magnetization reversal process, the switch from saturation into the vortex state, happens at the same external field (~ 20 kA/m) in both cases.

For the half-ball with hole, both curves are nearly identical. The oscillations which are visible near zero external field are again decreased by the larger value of α; however, both magnetization reversal steps – from positive saturation into the vortex state (the broad step in the curve) and further to negative saturation – occur at nearly the same external magnetic field.

Apparently, the influence of the value of α is largest for simulations in which strong precessions can occur, i.e. in samples which can contain a vortex core. In other samples, the differences caused by different values chosen for α can be very small or even negligible.

4.2.e. Influence of the field sweeping speed
Dimensions: Diameter 100 nm, height 50 nm, shape modifications
Material: Permalloy (Py)
Exchange constant A: 1.05 x 10^{-11} J/m
Magn. polarization at saturation J: 1 T
Gilbert damping constant α: 0.01
Tetrahedal mesh dimension: ≤ 3.7 nm
Field range: - 600 kA/m … + 600 kA/m (field along z-axis)
              - 450 kA/m … + 450 kA/m (field along x-axis)
Field sweeping speed: 1 kA/(m ns) / 0.1 kA/(m ns)

The shape of the simulated samples is identical to those used in Chapter 4.2.a, a graphical depiction is given in Fig. 4.2.16.

The field sweeping speeds used here are 1 kA/(m ns) and 0.1 kA/(m ns), the results of which are shown in Figs. 4.2.17 and 4.2.18. Additionally, the simulation from Chapter 4.2.a can be used to compare a field sweeping speed of 10 kA/(m ns) (cf. Fig. 4.2.2). In all three simulations, the external magnetic field is oriented along the x-axis.

Fig. 4.2.16: Simulated samples (a)-(f). For details of the shapes, see caption of Fig. 4.2.1.

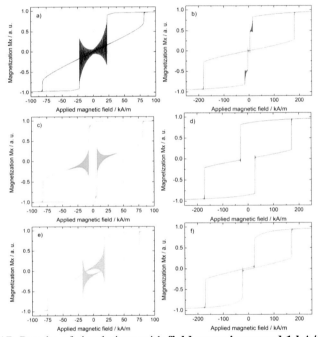

Fig. 4.2.17: Results of simulations with **field sweeping speed 1 kA/(m ns)** of the samples depicted in Fig. 4.2.16 for an external field along x.

Firstly, it is obvious that the oscillatory parts are "broader" for higher field sweeping speeds. The amplitudes of the oscillations are sometimes similar in all three cases; however, in some of the samples, different field sweeping speeds can lead to significantly different oscillatory amplitudes, e.g. for Figs. 4.2.17c and 4.2.18c.

The field values at which the different magnetization reversal processes – from saturation into a vortex state and further to opposite saturation – happen are nearly identical, if no oscillations occur in the respective field regions. As already recognized in Chapter 4.2.d, such oscillations can support the magnetization reversal process, leading to reduced switching fields.

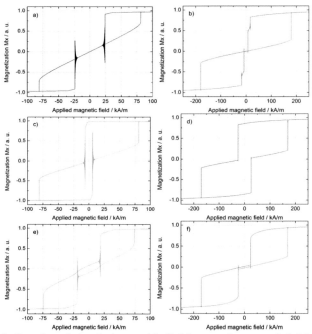

Fig. 4.2.18: Results of simulations with **field sweeping speed 0.1 kA/(m ns)** of the samples depicted in Fig. 4.2.16 for an external field along x.

Although the field sweeping speed can apparently change the simulation results, especially by introducing oscillations near "critical" field regions in which magnetization reversal processes occur, in most cases the simulated hysteresis loops are not significantly modified by a different choice of the field sweeping speed. Especially the coercive fields as well as the magnetization reversal mechanisms remain unaltered. Nevertheless, due to the possible strong influence of vortex core precessions on the magnetization reversal into the saturated state, high switching speeds can qualitatively and quantitatively alter the switching behavior of magnetic nano-dots in which a vortex core can occur. This effect has to be taken into account in the practical construction of half-ball-shaped nano-particles for storage and similar applications.

4.2.f. Magnetization reversal mechanisms in M_x-M_y-graphs

Dimensions: Diameter 100 nm, height 50 nm, shape modifications
Material: Permalloy (Py)
Exchange constant A: 1.05×10^{-11} J/m
Magn. polarization at saturation J: 1 T
Gilbert damping constant α: 0.01 / 0.1
Tetrahedal mesh dimension: ≤ 3.7 nm
Field range: - 600 kA/m … + 600 kA/m (field along z-axis)
               - 450 kA/m … + 450 kA/m (field along x-axis)
Field sweeping speed: 10 kA/(m ns)

In this chapter, different magnetic states and magnetization reversal mechanisms are examined by evaluation of the corresponding M_x-M_y-graphs. Two different Gilbert damping parameters α are used for the simulations as well as two different shapes, i.e. a solid half-ball (Fig. 4.2.19) and a half-ball with a cylindrical centered hole (Fig. 4.2.21).

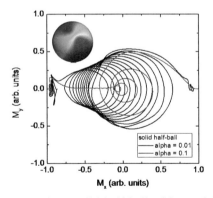

Fig. 4.2.19: M_y vs. M_x for a solid half-ball with α = 0.01 (black line) and α = 0.1 (red line).

In the solid half-ball, the helices are correlated to vortex core precession for both values of α,, i.e. they depict a vortex state. Near positive and negative saturation ($M_x = \pm 1$), the irregular changes in M_y are related to formation and vanishing of the vortex state. Opposite to the simulation for α = 0.01, for the larger value of α the precession ends with the vortex being centered relative to the sample, before the next magnetization reversal step into negative saturation occurs.

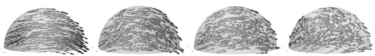

Fig. 4.2.20: Formation of the vortex state. After a "canting" of the magnetization in z-direction (top panel), the vortex core (here: red arrows) is formed in positive z-direction and starts precession, while parts of the magnetic moments (here: blue arrows) stay oriented in negative z-direction.

The formation of the vortex begins in both cases with a "canting" of the magnetization with opposite signs on opposite edges of the half-ball, so that the magnetization vectors can align partly along the upper rounded border of the half-ball (Fig. 4.2.20). At the same time, the magnetic moments are tilted into positive and negative y-direction on opposite sides of the half-balls. This tilting defines the vortex rotation direction and thus the vortex core direction ("upside" or "downside"). After an enhancement of the tilting, the connection between both halves of the magnetization (oriented in positive and negative z-direction) breaks up in an apparently chaotic process (3rd panel in Fig. 4.2.20), which leads to the formation of the vortex core and subsequently to the start of its precession (4th panel).

In the half-ball with symmetrical hole, no vortex core can occur, since the respective area is not available. Starting at positive saturation (Fig. 4.2.21 and Fig. 4.2.22, left panel), the magnetic moments are slightly canted to orient along the form of the ring, resulting in a horseshoe-like state (Fig. 4.2.22, 2nd panel).

The qualitatively different M_y values (Fig. 4.2.21) in the horseshoe area can be attributed to the different damping constants. Here, the vortex formation (Fig. 4.2.22, 3rd panel) is identical with $M_x \sim M_y \sim 0$. The decomposition of the vortex state into domains and the subsequent new ordering in the saturated state are accompanied by firstly smooth and afterwards fast changes in both magnetization components. Oppositely to the horseshoe-like state, the absolute values of M_x are > 0.5 in this region.

Apparently, such an M_x-M_y-graph can be used to examine qualitatively which magnetization states occur in a sample under examination, without checking all magnetization snapshots.

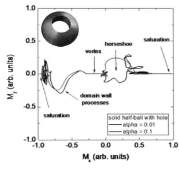

Fig. 4.2.21: M_y vs. M_x for a half-ball with cylindrical hole with $\alpha = 0.01$ (black line) and $\alpha = 0.1$ (red line).

In a similar way, M_x-M_y-graphs will be used for the next systems to examine whether similar conclusions can be drawn in fourfold or sixfold wire systems, where more different magnetic states are possible, depending on the system dimensions.

Fig. 4.2.22: Magnetization reversal in the half-ball with cylindrical hole. The colors of the arrows indicate the **z-direction** of the magnetization – red: positive z-direction; blue: negative z-direction. After a "canting" of the magnetization in z-direction during positive saturation (left panel), a horse-shoe state arises (2nd panel), followed by the flux-closed vortex state without any magnetization component in z-direction (3rd panel). Finally, the magnetization is reversed for negative saturation (4th panel).

4.2.g. Shape dependence of magnetization reversal
Dimensions: variable, see Table 4.2
Material: Permalloy (Py)
Exchange constant A: 1.05 x 10^{-11} J/m
Magn. polarization at saturation J: 1 T
Gilbert damping constant α: 0.01
Tetrahedal mesh dimension: ≤ 3.7 nm

Field range: - 600 kA/m … + 600 kA/m (field along z-axis)
- 450 kA/m … + 450 kA/m (field along x-axis)
Field sweeping speed: 1 kA/(m ns) and 10 kA/(m ns)

Table 4.1. Sample geometries.

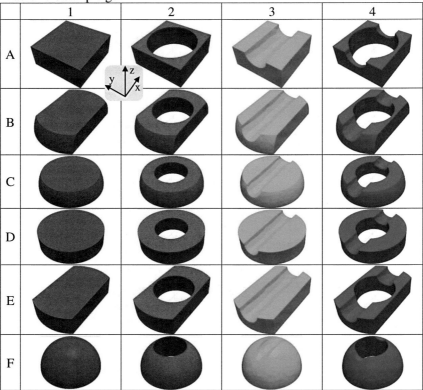

In the previous chapters, half-balls with different shape modifications have been examined. Now the simulations are expanded to 3D nano-objects with modifications of this form, not only by cutting parts but also by different corner shapes and partly diminutions of the samples in z-direction, taking into account several possibilities between the extrema of a cuboid and a half-ball.

Like in the previous chapters, again an additional cylindrical hole in the middle with diameter 25 nm and a cylindrical top-cut with diameter 10 nm have been taken into account, showing a first step in the transformation into a four-fold wire model.

An overview of the sample geometries in given in Table 4.1, the dimensions are listed in Table 4.2. It should be mentioned that the half-ball used here as sample F has only half the dimensions of the one examined before; this leads to different hysteresis loops.

Table 4.2. Sample dimensions. For the half-ball F, the sample height z is reduced by the further shape modifications (F2-F4).

	x / nm	y / nm	z / nm
A	28.5	28.5	12.5
B	50.0	28.5	12.5
C	50.0	50.0	12.5
D	50.0	50.0	12.5
E	50.0	28.0	12.5
F1	50.0	50.0	25

Interestingly, for A1 and A3, the z-direction is a hard axis, while it becomes an easy axis by cutting the cylindrical hole for samples A2 and A4. For the latter, the hysteresis loops for the magnetic field along the x-axis show a step around $H_{ext} = 0$ which corresponds to a flux-closed vortex state as seen in similar samples before (e.g. Fig. 4.2.22).

Opposite to the examination of the influence of the field sweeping speed in the half-ball (Chapter 4.2.e), the cubic sample simulated here offers not only quantitative, but also qualitative differences for both sweeping speeds. For A1 with the field along z, e.g., the reversible part of the hysteresis loop detected with 1 kA/(m ns) vanishes completely for a higher switching speed of 10 kA/(m ns).

For sample B, the z-direction always seems to be a hard direction. The hysteresis loops for B2 and B4 along z are nearly identical. The curves for B1 and B3 along z also differ only slightly. The field sweeping speed causes only small changes for all loops along the z-direction.

For a field sweep along the x-direction, sample B2 shows a loop with the broad step which is typical for a vortex state. This vortex state is introduced with strong oscillations, as has been seen before in the half-ball samples with cylindrical hole.

Opposite to A4, there is no such vortex state visible in B4. Apparently, here the cylindrical top cut has a stronger influence than in the larger sample A.

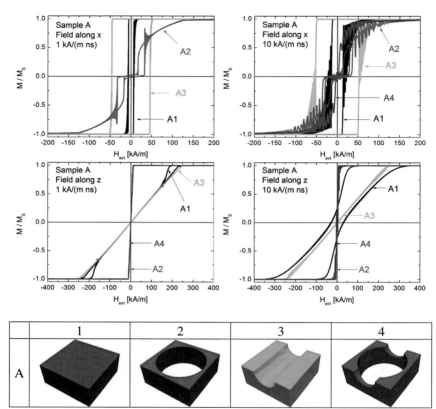

Fig. 4.2.23: Hysteresis loops for the four modifications of sample A for an external magnetic field along x (top panels) and z (bottom panels) and for field sweeping speeds of 1 kA/(m ns) (left panels) and 10 kA/(m ns) (right panels).

Sample C1 is the first with a rotational symmetry. Hysteresis loops for field sweeps along the x-direction are always typical easy-axis loops, independent of further shape modifications. Samples C1 and C2, however, show strong oscillations for a field sweeping speed of 10 kA/(m ns), which can apparently be significantly reduced by the horizontal cut in samles C3 and C4. Apart from these oscillations, the coercive fields and loop shapes do not differ strongly for both field sweeping speeds.

For a field sweep along the z-axis, similar to A1, sample C1 shows a deviation from the typical partly reversible hysteresis loops for the higher field sweeping speed of 10 kA/(m ns). All other sample modifications show very similar behavior for both field sweeping speeds.

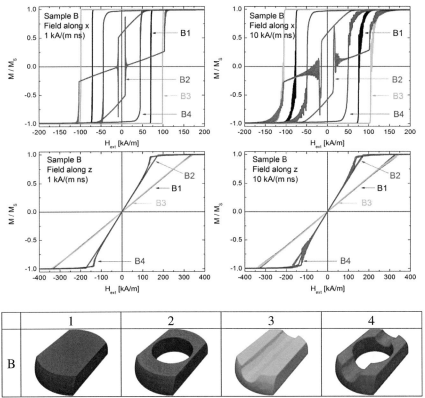

Fig. 4.2.24: Hysteresis loops for the four modifications of sample B for an external magnetic field along x (top panels) and z (bottom panels) and for field sweeping speeds of 1 kA/(m ns) (left panels) and 10 kA/(m ns) (right panels).

Sample D is nearly identical with sample C; however, the diminution of sample C in z-direction is removed here.

For the external field directed along x, both sets of samples result in nearly the same coercive fields and loops shapes, for field sweeping speeds of 1 kA/(m ns) as well as 10 kA/(m ns).

Sweeping the external magnetic field along the z-direction, the hysteresis loops are again quite similar to those simulated for sample set C. Only the jump from the irreversible part around $H_{ext} = 0$ into the reversible part for numerically larger values is noticeably shifted shifted for the higher field sweeping speed.

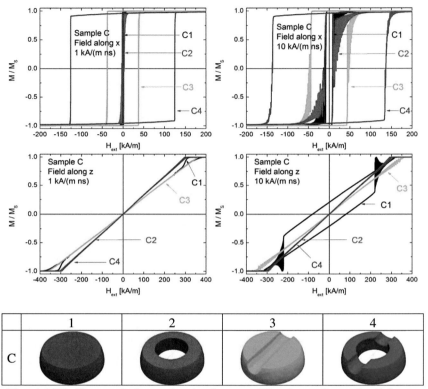

Fig. 4.2.25: Hysteresis loops for the four modifications of sample C for an external magnetic field along x (top panels) and z (bottom panels) and for field sweeping speeds of 1 kA/(m ns) (left panels) and 10 kA/(m ns) (right panels).

Sample E is similar to sample B, but the diminution of sample B in z-direction is removed here. For a field sweep along the z-direction, the hysteresis loops look very similar to those simulated for sample B, with nearly identical values of reaching positive and negative saturation for all four sample modifications.

Sweeping the external magnetic field along the x-direction also leads to similar coercive fields and loops shapes as in sample B for most of the sample modifications. In E2, however, a significant change is visible for the field sweeping speed of 10 kA/(m ns) – here the jump into the vortex state is skipped; instead after crossing the coercive field a reversible state is reached.

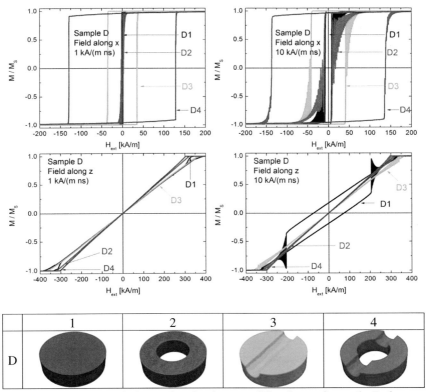

Fig. 4.2.26: Hysteresis loops for the four modifications of sample D for an external magnetic field along x (top panels) and z (bottom panels) and for field sweeping speeds of 1 kA/(m ns) (left panels) and 10 kA/(m ns) (right panels).

Sample F shows the well-known strong oscillations for field sweeps along the x-axis which have been recognized before (e.g. Fig. 4.2.15). However, opposite to a former comparison of the same field sweeping speeds for the same Gilbert damping constant of $\alpha = 0.01$, here a qualitative difference between the different field sweeping speeds can be recognized which can apparently be attributed to the reduced dimension (by a factor of 2): For 1 kA/(m ns), the typical vortex states with the corresponding broad steps vanish completely in all four sample modifications. This behavior is directly opposite to the finding in sample E2 where the higher field sweeping speed led to a magnetization reversal without vortex state.

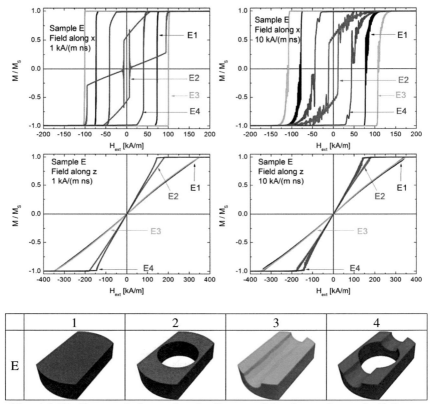

Fig. 4.2.27: Hysteresis loops for the four modifications of sample E for an external magnetic field along x (top panels) and z (bottom panels) and for field sweeping speeds of 1 kA/(m ns) (left panels) and 10 kA/(m ns) (right panels).

For an external magnetic field along the z-direction, the sweeping speed changes not only the field values which are necessary to reach saturation, but leads to a completely changed magnetization reversal process for sample F1 – for the higher field sweeping speed, the hysteresis loop shows the typical irreversible part around $H_{ext} = 0$, as it has already been recognized in both other samples with circular symmetry; for 1 kA/(m ns), it shows a 2-step magnetization reversal which has not been visible in any of the other samples before.

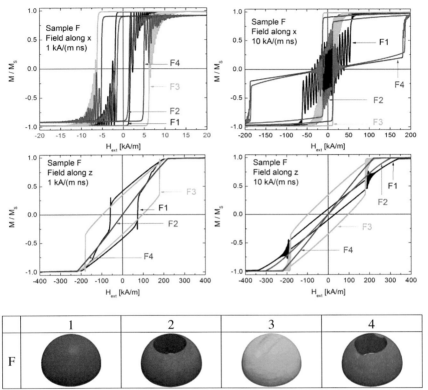

Fig. 4.2.28: Hysteresis loops for the four modifications of sample F for an external magnetic field along x (top panels) and z (bottom panels) and for field sweeping speeds of 1 kA/(m ns) (left panels) and 10 kA/(m ns) (right panels). In the first panel, the x-axis is different from the values used before.

This special situation is depicted in Fig. 4.2.29. For positive saturation (1st panel), all magnetic moments are aligned in positive z-direction, depicted as red arrows. The magnetization reversal starts, identical to the larger half-balls, with the formation of a vortex in the x-y-plane (2nd panel, green arrows). The vortex core is still directed along the positive z-direction. When the vortex core diameter is reduced with smaller external magnetic fields, the core starts a precession (3rd panel). Opposite to the larger half-ball, here the vortex core precession does not result in a flip of the core when the outer part of the half-ball has already finished the magnetization reversal, but an additional step becomes visible which is identical with the non-linear step in the hysteresis loop between ~ -70 kA/m and 140 kA/m: The vortex core moves to the border of the half-ball (4th panel), leading to the whole magnetization being oriented approximately in the x-y-plane and continuing the former precession of the vortex core with the complete magnetization of the whole sample. Within some collective precession rounds, the magnetization starts showing a growing tendency to be directed along the negative z-direction. However, the "canting" of the magnetization can still be seen, as it was already shown for a magnetization reversal along the x-direction in Fig. 4.2.20. Finally, with numerically larger negative fields, the external field dominates the form anisotropy, resulting in the final switch of the complete magnetization into negative saturation.

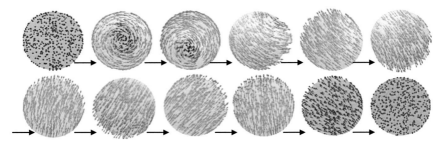

Fig. 4.2.29: Magnetization reversal of sample F1 for the external magnetic field swept along the z-direction with a speed of 1 kA/(m ns). Colors correspond to $M_z = 1$ (red) / 0 (green) / - 1 (blue).

The difference between the sample simulated here and the former results can be explained as follows: Since the particle used here is smaller about a factor of 2, the form anisotropy here has a larger influence, causing the difference between the energetically hard z-axis and the energetically favored x-y-plane to be of more importance. This leads to the situation that for not too large negative fields, the form anisotropy dominates, and the energetically favored magnetization orientation is a situation which actually belongs to the magnetization reversal in x-direction.

It should be mentioned that such a phenomenon, if it can also be found for higher field sweeping speeds, can also support the idea of more than two stable states at remanence – in this way, three stable states could be realized. However, since this effect has only been found for the lower field sweeping speed, is has not been further analyzed.

4.3. 4-fold wire systems

Nanopatterned magnetic structures are possible candidates for storage devices with higher data densities [Kar11]. Using the nanoporous template technology, high areal densities, exceeding the limits of conventional solution, can be enabled [Nie01, Fer99]. In such structures, a specific domain state of a single ferromagnetic object (particle) can define a single bit [Vil09]. Possible utilizations of magnetic states in nano-scale ferromagnetic systems are MRAM devices, magnetic logical circuits, or magnetic quantum cellular automata etc. [Jeo98, Jeo99, Cow00, Ake05, Ter05, Bad06, Bow09]. For these practical applications, flux-closed vortex states in nano-patterned rings are of special interest due to the minimization of stray fields and the occurrence of flux-closed vortex states [Cow99, Zhu00, Zhu04, He10, Zha10]. In such systems, different magnetic states can be represented by the chiralities (left or right) of the vortices, allowing for storage of binary digital information.

While the previous chapter concentrated on such systems with circular symmetry, now flux-closed and other magnetic states in fourfold systems are investigated, in order to find magnetic particles with more than two different magnetic states. Such systems would be of great interest for information storage devices, to enhance significantly the possible data density per unit area. A quaternary storage medium, e.g., with four distinguishable states (zero / one / two / three), would increase the number of storable information by a factore of two, since these four states could be treated as two binary bits, even without the need to change the usual digital logics. This idea has also inspired other authors who have reported on magnetic systems with three, four or even eight magnetization states in a variety of nano-structures [Wan09, Hua10, The10, Zha10, Mor11]. However, in most system these states cannot be detected easily, since the magnetization has to be measured at several distinct spatial positions to differentiate between similar states, what strongly enhances the experimental demands. This problem can, e.g., be solved for a system of coupled rings by breaking the symmetry, i.e. by coating some parts of them [Bow09]. Such an approach, however, complicates a technology and increases production costs.

Thus, patterned nano-magnets, produced, e.g., by nano-imprint lithography [Cro06], allowing for the detection of four or more states without the need of additional production steps, such as partial coating, are very interesting for utilization in magnetic storage devices and similar applications. This is why this chapter concentrates on stable intermediate states at remanence.

Parts of this chapter have been published in [Bla11] and [Bla13].

4.3.a. Hysteresis loops with intermediate states
Wire dimensions: Lengths 70 nm, diameters 10 nm
Wire orientation: 4 wires, 90° between neighbors, ends 5 nm free
Material: Iron (Fe)
Exchange constant A: 2×10^{-11} J/m
Magn. polarization at saturation J: 2.1 T
Gilbert damping constant α: 0.1
Tetrahedal mesh dimension: ≤ 3 nm
Field range: - 800 kA/m … + 800 kA/m (field in x-y-plane)
Field sweeping speed: 10 kA/(m ns)

Four iron (Fe) wires with diameter 10 nm and length 70 nm are oriented in a square configuration, the wires immersing near the ends (see inset in Fig. 4.3.1). Similar lateral dimensions (between 50 and 300 nm) are typically used in recent simulations of magnetic nano-particles [Dan10, He10, Li10, Mej10, The10, Tud10, Zha10a, Mor11]. Afterwards, the system has been scaled down to smaller dimensions which are comparable to recent grain or dot sizes. While recent hard disks are based on perpendicular recording, the system described here is ordered in-plane, as usual for nano-patterned rings [Cow99, Zhu00, Zhu04, He10, Zha10] etc., to support the comparison between both kinds of magnetic nano-structures. The simulated systems can be defined by wires with a given length-to-diameter aspect ratio and four local crossing regions. These features are responsible for the subsequent demagnetizing fields driving the particle.

To avoid contributions of neighboring nano-magnets, a single magnetic particle has been simulated to verify the existence of four stable magnetic states in principle, as often practiced in actual publications about magnetic nanosystems [Dan10, He10, Hua10, The10, Zha10, Li10]. The influence of magnetic stray fields of neighboring particles, depending on distance and respective orientation, will be simulated in future to optimize the system for application in magnetic storage media.

In the simulation, finite tetrahedal elements of maximum 3 nm diameter were used for meshing, which is smaller than the Néel exchange length of 3.69 nm [Sch03a]. The other physical parameters were chosen as follows: exchange constant $A = 2 \cdot 10^{-11}$ J/m, magnetic polarization at saturation $J_s =$ 2.1 T, and the Gilbert damping constant $\alpha = 0.1$ [Kne91].

The external magnetic field was applied in the sample plane, for different sample orientations starting with that parallel to the x-axis (0°) which is parallel to one pair of wires. The field vector was always kept parallel to the x-axis. Simulations begun from a random magnetization state (external magnetic field H_{ext} = 0); then the field was swept at a constant speed of 10 kA/(m·ns) up to 800 kA/m to saturate the sample. Afterwards, a complete hysteresis loop to -800 kA/m and back to positive saturation at + 800 kA/m has been performed. Modifications of this procedure are mentioned in the respective figure captions.

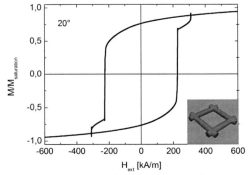

Fig. 4.3.1. Hysteresis loop, simulated for the nano-wire system shown in the inset, for 20° between one pair of wires and the external magnetic field. From [Bla11].

In Fig. 4.3.1, a hysteresis loop is depicted, simulated for the sample oriented at 20° with respect to the x-axis and the external magnetic field. Both sides of the loop show steps, implying a two-step magnetization reversal process as it is known from other magnetic objects, like e.g., nano-rings [Bow09, Zha10], but also from some exchange-biased systems [Til06].

In order to verify if these steps can be attributed to *stable* intermediate states, which could be used in magnetic storage media, the following magnetic field sequence is used for the next simulations: Starting at positive saturation, the external field is swept to the region of the first intermediate step and afterwards back to zero field. Afterwards, the first half of the hysteresis loop is finished to negative saturation. The second half of the loop is simulated in the same way, via the second intermediate step region and zero-field, with reversed sign of the external field. The whole procedure is shown in Fig. 4.3.2(a). The resulting curves are depicted in Fig. 4.3.2 (b)-(f) for various sample orientations. While the intermediate step is relatively flat and broad for smaller angles, it becomes higher and narrower for larger angles. A sample angle of about 20° seems to be ideal for the detection of all different states, since here the four magnetic states are optimally separated for vanishing external field. Obviously, for this angular orientation, measuring the net magnetizations would result in all four states being detected. This is a large advantage of the proposed solution in comparison with vortex states in symmetric nano-rings – measuring on one or more exact spatial positions within the particles is not necessary here.

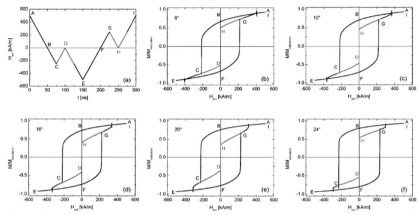

Fig. 4.3.2. (a) Field sweeping scheme for the hysteresis loops shown in (b)-(f) for different angles between the nano-wire system under examination and the external field. From [Bla11].

To compare the nano-wire system with the other systems with similar step-like behavior in the hysteresis loop, i.e. nano-rings and exchange-biased bilayers like Fe/MnF$_2$ [Til06], snapshots of the spatial distribution of the magnetization in the system are depicted in Fig. 4.3.3 for prominent positions (see Fig. 4.3.2(a) for the definition of the steps A-I).

Fig. 4.3.3. Magnetization reversal mechanism in nano-wire system for positions of the four different states at vanishing external magnetic field, as depicted in Fig. 4.3.2(a).

All stable states at remanence (B) are onion states. Opposite to nano-rings, no vortex-like-state can be found here. Instead, the system behaves similarly to the exchange-bias system Fe/MnF$_2$ with twinned antiferromagnet [Til06], in which the magnetization shows coherent magnetization rotation to a stable intermediate state oriented at about 90° to the saturation magnetization orientations.

For a better understanding of scaling effects in the system under investigation, simulations have also been performed with systems of smaller dimensions. These sizes are closer to typical actual grain sizes of about 10 nm or to dot sizes of about 20 nm which are necessary to reach area densities of 1 Tbit/inch² if conventional binary storage is used [Mor11].

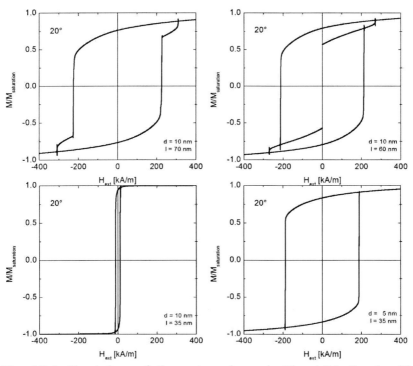

Fig. 4.3.4. Simulations of the system shown in Fig. 4.2.1 (Inset) with different dimensions. From [Bla11].

Fig. 4.3.4 shows systems with different wire lengths and diameters for a sample orientation of 20°, which was found to be ideal in the original system. Reducing the wire length to 60 nm leads to narrower and less broad steps. If a wire length of 35 nm (half the original length) is simulated, the whole system reverses magnetization as one single domain; the hysteresis loop is much narrower. The hysteresis becomes broader again if also the wire diameter is reduced; however, the step is no longer visible.

As this short outlook has shown, downscaling the system leads to qualitatively different magnetization reversal mechanisms, which may necessitate changing the material (e.g. to get a smaller exchange length etc.) in order to gain systems with four stable magnetic states at remanence again. A more detailed overview of scaling effects in fourfold ferromagnetic systems will be given in Chapter 4.3.c. Such scaling effects have to be taken into account for utilization in magnetic storage applications, e.g. by changing the material or adjusting the shape.

Apparently, fourfold nano-wire system offer the possibility to create a patterned magnetic system with four unambiguously distinguishable states in vanishing external field, which can thus be utilized for quaternary storage devices (four states / two bits per particle). Opposite to nano-rings, no vortices are used during magnetic-based data processing; instead, the intermediate states show global net magnetizations.

Further micromagnetic simulations, e.g. on sixfold samples (Chapter 4.6), will show whether even more than four stable states can be achieved by respectively constructed patterned nano-wire systems.

4.3.b. Angular dependence of the coercive fields – 2x2 wires
Wire dimensions: Lengths 70 nm, diameters 10 nm
Wire orientation: 4 wires, different distances
Material: Iron (Fe)
Exchange constant A: 2 x 10^{-11} J/m
Magn. polarization at saturation J: 2.1 T
Gilbert damping constant α: 0.01
Tetrahedal mesh dimension: ≤ 3 nm
Field range: - 800 kA/m … + 800 kA/m (field in x-y-plane)
Field sweeping speed: 10 kA/(m ns)

Structures composed of magnetic nanowires have been examined for utilization in diverse novel magnetic systems, such as magnetic logical circuits or magnetic quantum cellular automata [Cow02, Don08, Pul10]. In such nanowire systems, as well as in other patterned magnetic structures, the coupling between neighboring magnetic units is of great importance for the magnetization reversal process of the complete system [Hen01, Váz04, Sil06, Kar11]. For coupled nano-rings, e.g., the overlap of neighbouring rings has shown a strong influence on the signal transport properties in logical NOT gates etc. [Bow09], while in some systems a non-negligible probability of unintentional signal inversions due to random effects was observed [Wel03, Wel03a].

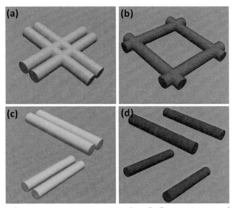

Fig. 4.3.5: Magnetic systems composed of four nano-wires, arranged in perpendicular pairs, with different coupling within one pair / between the pairs: Maximum coupling between the pairs and (a) coupling / (b) no coupling within each pair; no coupling between the pairs and (c) coupling / (d) no coupling within each pair.

While the behaviour of single ferromagnetic wires is well understood, simply governed by the shape anisotropy, the interplay between nano-wires of different orientations and coupling requires elementary studies [Her01, Pit11]. Since most commercial magnetic solutions nowadays are based on thin-layers technology, where the coupling between different materials can tailor these devices' functionalities via surface or bulk interactions (2D or 3D; surface or bulk), an approach is analyzed which uses coupling between parallel wires (1D coupling) and/or perpendicularly oriented ones (0D coupling). The results of magnetization dynamics point out new functionalities of these novel wire-sets. Different coupling mechanisms can significantly influence the magnetic anisotropies, as will be shown by angular-dependent simulations of the coercive fields of four different fourfold samples. In this way, the results presented here broaden the knowledge about shape, size and composition analysis of nano-wire magnetism [Her09, Fon11].

For the simulations, four different systems with the following coupling possibilities were taken into account (Fig. 4.3.5): maximum coupling between the pairs and coupling / no coupling within each pair; no coupling between the pairs and coupling / no coupling within each pair. The coercivities of these four magnetic nano-systems have been derived from the simulated hysteresis loops.

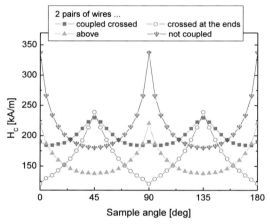

Fig. 4.3.6: Simulated coercivities H_C of the magnetic systems depicted in Fig. 4.3.5.

Figure 4.3.6 shows the calculated values of H_C for the angular region of 0° ... 180°. As expected in systems with fourfold geometry, a fourfold anisotropy can be detected in all four situations. From former simulations [Til09] it is known that a "typical" four-fold anisotropy leads to sharp maxima of the coercivity along the easy axes and broad minima around the hard axes, as can be seen here for the green and the blue curve, i.e. no coupling between the two pairs. However, the other situations show deviations from this behavior. Assuming that the identification of easy axes by maxima in the coercivity is universal and can thus be used for all cases under examination, even an exchange of easy and hard axes of systems, as resulting from different couplings between the pairs of nano-wires, can be recognized.

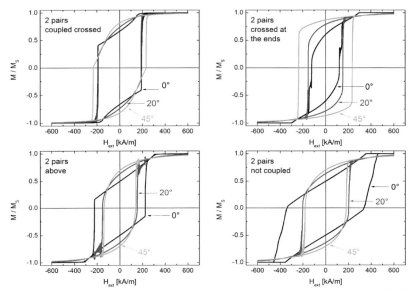

Fig. 4.3.7: Simulated hysteresis curves for the samples depicted in Fig. 4.3.5, exemplarily shown for orientations relative to the external magnetic field of 0° (i.e. parallel to one pair of wires), 20° and 45°.

For a deeper understanding of this finding, the simulated hysteresis loops are depicted in Fig. 4.3.7 for orientations relative to the external magnetic field of 0° (i.e. parallel to one pair of wires), 20° and 45°. In all four cases, the simulated loops show deviations from "typical" hysteresis curves. Mostly, the shapes can be regarded as composed of two hysteresis loops, one with typical attributes of an easy axis (i.e. broad loop with abrupt magnetization changes) and one with typical signs of a hard direction (i.e. narrow loop with broad transition regions from one saturated state to the other). Apparently, easy and hard axes cannot easily be defined in the wire systems under examination.

Comparing Fig. 4.3.6 – with absolute maxima in the coercivity around 0° for cases (c) and (d) – and Fig. 4.3.7 – with abrupt magnetization changes at 0° for cases (a) and (d) –, it is obvious that the identification of an easy fourfold axis by a maximum in the coercivity is no longer valid in the systems under examination. Instead, the coercivities are related with the magnetization reversal processes which may differ in dependence of the coupling configuration and the angle.

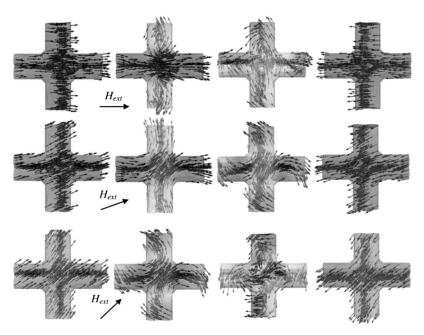

Fig. 4.3.8: Magnetization reversal of the two pairs **coupled crossed** from positive to negative saturation for sample angles of 0° (top row), 20° (middle row), and 45° (bottom row). The colors depict a magnetization orientation in positive (red) or negative (blue) x-direction; the external field direction is marked in each row.

To examine these magnetization reversal processes in detail, Figs. 4.3.8-4.3.11 show snapshots of the magnetization between positive and negative saturation for the four wires systems depicted in Fig. 4.3.5, each for the sample angles 0°, 20°, and 45°, the hysteresis loops for which are depicted in Fig. 4.3.7.

For the coupled crossed sample (Fig. 4.3.8), in the sample orientation 0° the magnetization reversal process occurs via a "mixed" state in which the "horizontal" wires (parallel to the external field) keep the positive saturation, while the wires perpendicular to the field show an onion-like state (2nd panel). The further magnetization reversal happens via domain wall processes.

For the sample orientation of 20°, an x-like state (2nd panel) is visible, with the magnetization in the crossing region being oriented approximately diagonally. This state is deformed into a double-C state (3rd panel), leading to negative saturation without domain wall processes.

For the sample orientation of 45°, the double-C state can also be recognized; however, the final magnetization reversal occurs via a domain wall process.

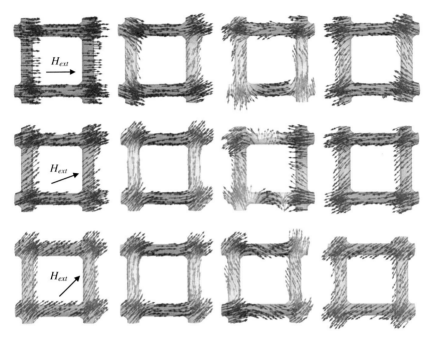

Fig. 4.3.9: Magnetization reversal of the two pairs **crossed** at the ends from positive to negative saturation for sample angles of 0° (top row), 20° (middle row), and 45° (bottom row). The colors depict a magnetization orientation in positive (red) or negative (blue) x-direction; the external field direction is marked in each row.

In the crossed pair of samples, for 0° the reversal process occurs via two different horse-shoe states (2nd and 4th panel) and a flux-closed vortex state.

In the 20° orientation, an onion state (2nd panel) is visible, followed by a magnetization reversal via domain wall processes.

For 45°, the onion state (2nd panel) is followed by an "extreme onion" (3rd panel) in which the magnetization is no longer parallel to the wire orientation but starts to reverse at all positions at the same time.

Fig. 4.3.10: Magnetization reversal of the two pairs **above** from positive to negative saturation for sample angles of 0° (top row), 20° (middle row), and 45° (bottom row). The colors depict a magnetization orientation in positive (red) or negative (blue) x-direction; the external field direction is marked in each row. The apparent deviation from a fourfold symmetry results from the perspective depiction of the two pairs which are not positioned in the same plane (see Fig. 4.3.5).

In the sample with the wires above each other, the coupling between both pairs is negligible. Here, the magnetization reversal starts for a sample orientation of 0° with a reversal in the wire pair perpendicular to the magnetic field, since here it is supported by the form anisotropy. The magnetization direction in both wires is opposite to reduce stray fields (2nd panel). The reversal process in the wires parallel to the external field works via domain wall processes.

Fig. 4.3.11: Magnetization reversal of the two pairs **non-coupled** from positive to negative saturation for sample angles of 0° (top row), 20° (middle row), and 45° (bottom row). The colors depict a magnetization orientation in positive (red) or negative (blue) x-direction; the external field direction is marked in each row. The apparent deviation from a fourfold symmetry results from the perspective depiction of the two pairs which are not positioned in the same plane (see Fig. 4.3.5).

For 20°, the whole process is similar; however, the magnetization reversal in the wires perpendicular to the magnetic field now is dominated by the external field instead of the form anisotropy, resulting in identical magnetization directions in both wires. At a sample orientation of 45°, the reversal starts by a magnetization rotation in both pairs of wires, leading to negative saturation again via domain wall processes.

In the non-coupled pair of wires, the magnetization reversal starts again in the wires perpendicular to the external field. Both other wires change the magnetization direction via domain wall processes at independent magnetic fields (3rd panel). Opposite to the wires coupled at the end, there is no collective phenomenon like a vortex state.

A similar process can be seen for a sample orientation of 20°; however, here no domain wall processes are included.

For 45°, after a similar start of the reversal process, the last step includes again different domain wall processes.

The micromagnetic simulations of systems consisting of two perpendicular pairs of parallel wires have shown the strong dependence of the magnetic properties on the strength of coupling between the pairs and within them. Additionally, some systems exhibit unexpected features, esp. a step in the hysteresis loop, which are promising for the development of new functionalities.

4.3.c. Influence of dimensions on magnetization reversal processes
Wire dimensions: Lengths 30 nm - 70 nm, diameters 6 nm - 20 nm
Wire orientation: 4 wires, crossed at the ends
Material: Iron (Fe)
Exchange constant A: 2×10^{-11} J/m
Magn. polarization at saturation J: 2.1 T
Gilbert damping constant α: 0.1
Tetrahedal mesh dimension: ≤ 3 nm
Field range: - 600 kA/m … + 600 kA/m (field in x-y-plane)
Field sweeping speed: 10 kA/(m ns)

In recent articles, some exotic magnetic states in wire-arrays and single objects are reported. Subrami *et al.* examined stable onion states in Permalloy (Py) rectangular rings of size 1.5 μm x 1 μm, with a cross-section of 250 x 250 nm^2 [Sub04]. Rectangular arrays of 1.15 μm x 0.7 μm rings with a closely (100 nm) and a widely (500 nm) spaced option for packing were investigated by Wang *et al.* [Wan05], who concluded that closer spacing influences transitions from vortex states into onion states and vice versa, due to more collective behavior of rings. Remhof *et al.* [Rem08] as well as Westphalen *et al.* [Wes08] studied closely spaced squared Py dipole lattice systems of four-fold symmetry and iron Kagome lattices (a special composition of interlaced triangles) of six-fold symmetry, interacting via dipolar fields. The observed collective effects were interpreted as occurrence of horseshoe and vortex states due to interactions between parallel and anti-parallel magnetically ordered sublattices.

A large set of magnetic states was also reported by Gao *et al.* [Gao07] who observed different routes for magnetization evolution including transitions between a domain wall, diagonal onion, horseshoe and a transition leading to vortexes in deformed cobalt (Co) rings. Those rings with 880 nm in length along the major axis and 660 nm along the minor axis had a ring width of 300 nm and an inter-ring edge-to-edge spacing of 500 nm. The difference between easy and hard directions for a reversal analysis was interpreted to result mostly from the dynamics of a single object in a relatively far spaced array, when the influence of magnetic stray fields was less significant. Similarly, He *et al.* searched for novel magnetic states in single magnetic objects, utilizing a slotted Co nanoring [He10]. The slot was created to break closed vortex flux, analysis was only carried out for the two main orientations of the externally applied magnetic field equivalently to a hard and an easy directions. A magnetic nanocylinder was studied by W. Zang and S. Haas by means of Monte Carlo analysis of the switching behavior influenced by shape modifications [Zha10]. Yoon *et al.* recently examined tail-to-tail domain wall motion in a circular ring made from a wire to reveal a spin-valve behavior [Yoo12].

The diameters of tested wires in a square configuration, superimposed at the ends, are chosen between 6 nm and 20 nm and the lengths between 30 nm and 70 nm, corresponding to length-to-diameter ratios between ~ 3 and 11. The length/diameter aspect ratio of the wires determined the volume of crossing regions and mainly influenced the resulting demagnetizing fields. Simulations were carried out for several angular in-plane directions of externally applied field, ranging from 0° to 45°.

It should be mentioned that, oppositely to the half-ball samples examined before, no oscillations of the magnetization vector were observed in the presented wire samples.

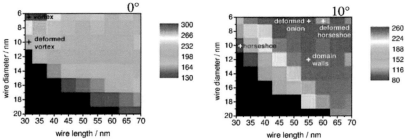

Fig. 4.3.12: Color-coded plot of coercivities, simulated for different orientations of externally applied magnetic field in (x-y) plane – the angle measured between the magnetic field intensity vector and the x-axis: 0° with the position of vortex and disturbed vortex, 10° with the positions of coherent magnetization reversal and dominating horseshoe states, via deformed horseshoe, via deformed onion state, and with domain-wall process and intermediate states excluded marked by crosses, respectively. The coercivity scale is expressed in (kA/m) units. Black regions indicate non-simulated configurations. From [Bla13b], modified.

Fig. 4.3.12 depicts the results of the simulations. The color code gives the coercive fields in (kA/m). Comparing the graphs for the different orientations, it is obvious that 0° and 10° show quite different results. For 0° (and 45°, not shown here), the largest coercivities can be found for systems with short and thin wires, and the smallest coercivities occur for systems of wires with larger diameters, almost independent of the wire length (different colors are roughly oriented horizontally).

Changing the field orientations from 0° to 10° (and from 45° to 40°, not shown here) modifies the coercivity characteristics significantly making comparable coercivity regions oriented more diagonally.

In order to test numerical stability and to avoid unintentional artifacts in hysteresis loops, 0° and 45° results have been compared with those obtained for slightly tilted directions of the applied field, that is, for 1° and 44° orientations, respectively. The results indicate that 0° is an instable direction in the simulation, changing the effects found here very fast for directly neighboring directions for all tested samples. On the other hand, the 44° result shows a qualitatively similar tendency only for samples of relatively small dimensions. In that sense the 45° direction is a stable direction in the simulation.

The strong quantitative differences of the coercivities for different wire lengths and diameters lead to the idea that different magnetization reversal processes may occur for different orientations and dimensions of the wire systems. Indeed, the magnetization snapshot of the simulations show several different magnetization reversal mechanisms, which are indicated exemplarily in both panels of Fig. 4.3.12, i.e. undisturbed vortex states, vortex state deformed by an out-of-plane component, coherent magnetization reversal, horseshoe and deformed horseshoe states, deformed onion states, and domain wall nucleation and propagation processes.

Examples of all magnetization states are given in Figs. 4.3.13-4.3.14. The graphs depict snapshots of different magnetization reversal processes which have been found in the samples under simulation, together with the hysteresis loop of the respective systems.

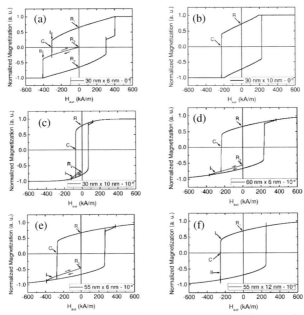

Fig. 4.3.13: Hysteresis loops of the following systems: 30 nm x 6 nm (length x diameter) system with 0° orientation (a); 30 nm x 10 nm system with 0° orientation (b); 30 nm x 10 nm system with 10° orientation (c); 60 nm x 6 nm system with 10° orientation (d); 55 nm x 6 nm system with 10° orientation (e); and 55 nm x 12 nm system at 10° (f). From [Bla13b], modified.

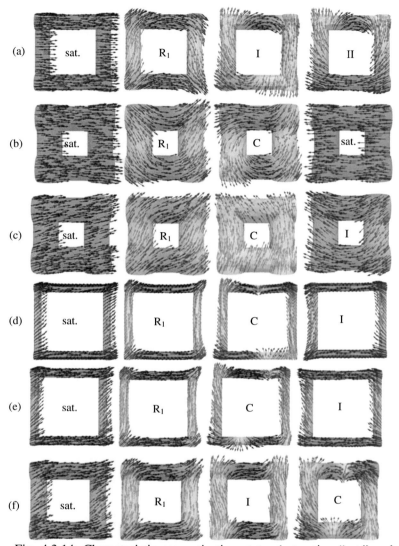

Fig. 4.3.14: Characteristic magnetization states (saturation "sat." and states correlated to letters in Fig. 4.2.21) in the magnetization reversal processes depicted in Figs. 4.2.21 and 4.2.22.

Figs. 4.3.13a and 4.3.13b show different possibilities of magnetization reversal for a system orientation adjusted to 0° (i.e. with one pair of wires parallel to the external magnetic field), via a vortex state (state marked as II in (a), leading to a step in the hysteresis loop (an intermediate state), or by a similar state with an out-of-plane magnetization component (marked as C in (b)), leading to a hysteresis loop without steps, respectively.

The additional four processes have been also found in simulated systems rotated away from 0°: the coherent magnetization reversal via horseshoe states and leading to the same horseshoe at stable intermediate state (marked as R_2 in (c)), only occurring in small particles with relatively thick wires; the reversal via deformed horseshoe states and domain walls at coercivity field (C) leading to a stable intermediate state (R_2) as a horseshoe one (d); next, the similar reversal via deformed onion states and domain walls at coercivity field (C) leading to a stable intermediate state (R_2) as an onion one (e), which are often seen in larger systems with relatively thinner wires; and finally, the reversal via domain walls nucleated and propagated along wires observable in particles of all dimensions (f).

The reversal processes leading to stable intermediate states – i.e. coherent rotation, horseshoe, and onion state – can be recognized by steps in the hysteresis loops (in some cases not very broad or with a magnetization component M_x quite near to the saturation magnetization) as it can be seen in (c)-(e). No step occurs in reversal with vortex disturbed by a domain wall (b) and domain wall dominated processes (f).

It should be mentioned that in some samples the magnetization reversal process is performed by more than one of the above described processes, sometimes leading to two separate steps on either side of a hysteresis loop or other effects which are not described here in detail.

Apparently, the fourfold wire systems under examination exhibit six different types of reversal processes, depending on system dimensions and orientation to the external magnetic field. For reversal processes along intermediate states, the corresponding hysteresis loops exhibit a step. For highly symmetric orientations, i.e. 0° and 45°, the dimensional dependence of the coercive fields – diameter vs. length diagrams – is nearly opposite to that in less symmetric orientations. The less symmetric orientations are obviously more susceptible to magnetization reversal via intermediate states. The susceptibility for such reversal in magnetic wire nano-structures is possible for a rich set of samples dimensions visible as separate regions derived from coercivity characteristics. Taking advantage from the competition between exchange and demagnetizing fields at nanoscale, it was shown that this effect can be tailored by a shape modification (length-to-diameter-ratio) indicating possible applications in next generation memory devices based on patterned structures.

4.3.d. Magnetization reversal mechanisms in M_x-M_y-graphs
Wire dimensions: Lengths 70 nm, diameters 10 nm
Wire orientation: 4 wires, crossed at the ends
Material: Iron (Fe)
Exchange constant A: 2 x 10^{-11} J/m
Magn. polarization at saturation J: 2.1 T
Gilbert damping constant α: 0.01
Tetrahedal mesh dimension: ≤ 3 nm
Field range: - 800 kA/m … + 800 kA/m (field in x-y-plane)
Field sweeping speed: 10 kA/(m ns)

Similar to the half-ball samples, the simulations of the four-fold wire samples can also be depicted in form of M_y vs. M_x graphs (Fig. 4.3.15).

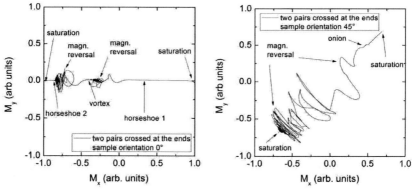

Fig. 4.3.15: M_y vs. M_x for the four-fold wire sample constructed from two pairs crossed at the ends (cf. Chapter 4.3.b), depicted for sample orientations of 0° (left panel) and 45° (right panel).

For a sample orientation of 0° (left panel), the magnetization reversal steps with the subsequent oscillations can be identified quite well. Comparing M_x and M_y components along the magnetization reversal process, two horseshoe states and a vortex state can also be recognized.

The magnetization reversal for a 45° orientation (right panel) is accompanied by stronger oscillations, i.e. by larger deviations from a linear connection between positive and negative saturation.

Although these graphs allow for a certain insight into the magnetization reversal process, it should be mentioned that the magnetization snapshots of these processes (cf. Fig. 4.3.9) offer much more information. While M_y vs. M_x graphs can be very helpful to understand, e.g., whether coherent rotation of the magnetization occurs [Bec03], they are of minor benefit in understanding magnetization reversal processes in magnetic nano-systems in detail.

4.3.e. Influence of wire connections
Wire dimensions: Length 70 nm, diameters 10 nm
Wire orientation: 4 perpendicular wires, different connections
Material: Iron (Fe)
Exchange constant A: 2 x 10^{-11} J/m
Magn. polarization at saturation J: 2.1 T
Gilbert damping constant α: 0.1
Tetrahedal mesh dimension: ≤ 3 nm
Field range: - 600 kA/m … + 600 kA/m (field in x-y-plane)
Field sweeping speed: 10 kA/(m ns)

Additional to the four-fold wire systems examined in Chapters 4.3.b and 4.3.c, another system with half-balls as corner connections has been simulated. The extended corners have been chosen for a better comparability with experiments on similar nano- or micro-systems, since nano-structuring can result in rounded corners.

For a direct comparison of three different corner solutions – crossed near the wire ends, overlapping at the wire ends, and extended with half-balls – a sample orientation of 16° has been chosen, since this orientation resulted in pronounced steps for the sample crossed near the ends (cf. Fig. 4.3.2).

Fig. 4.3.16 shows simulated hysteresis loops for the three different sample geometries depicted in the insets. Obviously, shifting the crossing point to the wire ends results in vanishing of the step in the loop. Additional half-balls, however, change not only the coercive fields significantly, but do also support the formation of the step during magnetization reversal.

Apparently, not only the length-do-diameter-ratio, but also the "corner part" – in which the form anisotropy is changed, compared to a single wire – of the samples decides about the magnetization reversal processes. If the part between the corners becomes smaller, compared to the corner area, new magnetization reversal processes occur, leading to the examined step.

It should be mentioned that the wire system with half-balls at the corners consists of half-wires, not of full cylinders. Even this major difference to the system with the wires crossed near the ends (black inset in Fig. 4.3.16) does not result in a change of the slope of the loop, while shifting the crossing points (green inset) leads to a qualitative difference of the hysteresis shape.

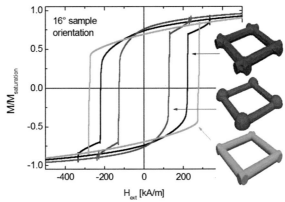

Fig. 4.3.16: Hysteresis loops for four-fold wire samples with different corner solutions. Insets: geometries of the three samples.

For a more detailed examination of the influence of the connection between the perpendicular wires at the corners, a set of samples has been produced with the ends crossed in the starting simulation and more and more shifted to the outside until only a point-like contact is left (Fig. 4.3.17).

Fig. 4.3.17: Five samples with the wire ends shifted about 0 nm (left), 5 nm, 8 nm, 9 nm, and 10 nm (right).

Fig. 4.3.18 shows the resulting angle-dependent coercivities. While stronger contacts between the perpendicular wires (black and red system) lead to a maximum in the coercivity for a sample orientation of 45° with respect to the external field, this maximum splits for reduced contacts and results in a minimum at 45°.

This finding can explain the results shown in Fig. 4.3.6 in which an exchange of hard and easy axes between coupled and non-coupled wires could be found. The new results depict that this qualitative change of the anisotropy axes is a continuous process instead of a sudden exchange at a certain amount of contact.

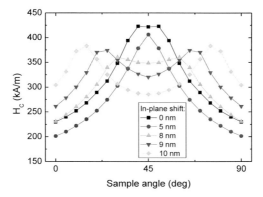

Fig. 4.3.18: Angle-dependent coercivities for the samples depicted in Fig. 4.3.17.

4.3.f. Special corner solutions

Due to the strong influence of the corner form on the anisotropies of a sample, different corner constructions have been examined to check whether one of them might be advantageous for the creation of more than two stable states at remanence. Fig. 4.3.19 shows the results of four different corner solutions for a sample orientation of 20° with respect to the external field: the usual direct connection without additional magnetic part or, oppositely, cuts (upper left panel); corners with additional balls (upper right panel); corners with balls which include quarter cuts (lower left panel); and finally corners with cuts of half the wire width (lower right panel).

Comparing the respective hysteresis curves, the ball-shaped corners lead to more possible states at remanence than the simple square; however, as already indicated by the several steps in the hysteresis loops, each single part of the system (wires at 20° to the external field, wires at 70° to the external field, different balls …) changes magnetization at a different field, leading to a nearly chaotic behavior. Since the last system shows a more even hysteresis curve, it can be expected to be of more interest for a reliable magnetization reversal process with a few distinct steps.

Thus, this system has been examined for different dimensions. Firstly, due to the requirements in the practical realization of such systems by lithography (much thicker samples result in problems of the photo-resist), all samples have been simulated with a thickness of 10 nm, while the length and the widths of the "walls" have been varied.

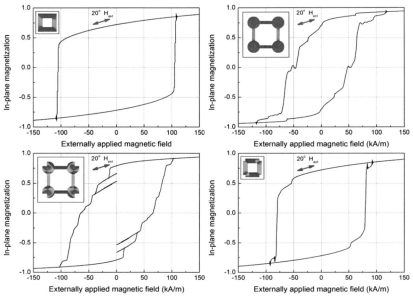

Fig. 4.3.19: Samples with different corner solutions (see insets) and respective coercive loops at a sample orientation of 20°.

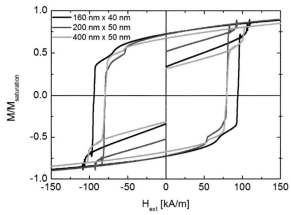

Fig. 4.3.20: Comparison of samples with thickness 10 nm and different dimensions, simulated at a sample orientation of 20° with respect to the external field.

Fig. 4.3.20 gives an overview of the results. In all three cases, a step in the hysteresis loop is visible, connected with a stable intermediate state, as verified by the additional field sweeps back to remanence. While increasing the system dimensions proportionally from 160 nm "wall" length and 40 nm "wall" width to 200 nm x 50 nm results in a less pronounced step, increasing only the "wall" length to 400 nm x 50 nm leads to a higher and longer step (green curve). Apparently, the square cuts in the corners enable relatively free scaling of the system without losing the stable intermediate states necessary for a quaternary storage system – something which was impossible in other systems (cf. Fig. 4.3.4).

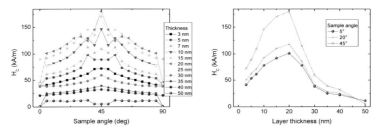

Fig. 4.3.21: Angle dependent coercivities for different sample thicknesses (left panel) and thickness dependent coercivities for some sample angles.

Additionally to the lateral dimensions, the system height has been varied. Fig. 4.3.21 shows the angle dependence of systems with lateral dimension 160 nm x 40 nm and with different heights.

While for layer thicknesses up to 40 nm, a maximum in the coercivity can be found around 45°, partly split into two neighboring maxima, the thickest sample shows nearly no angle dependence. Apparently, in the thicker samples, the anisotropies due to the lateral geometry become less dominant.

For all sample angles, a maximum coercivity can be found at a layer thickness of 20 nm (Fig. 4.3.21, right panel). Due to this finding, the hysteresis loops for a sample orientation of 20° with respect to the external field are depicted in Fig. 4.3.22.

Interestingly, for thicker samples, the step in the loop is significantly increased. Such a broad step as depicted here for a layer thickness of 30 nm can strongly support the technical possibilities of reading/writing. Thus, future research will, amongst others, concentrate on finding the ideal sample dimensions which can be structured by lithography without technical problems and, at the same time, lead to broad, pronounced steps.

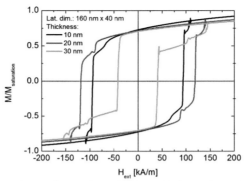

Fig. 4.3.22: Comparison of samples with lateral dimensions 160 nm x 40 nm and different thicknesses.

4.3.g. Magnetization reversal in fourfold Co wire system

While the previous simulations have been performed on Fe samples, this chapter shows results of simulations of a fourfold wire system made from Co. Thus, some of the parameters in the list below are changed compared to the other chapters (marked with bold letters): The exchange constant A as well as the magnetic polarization at saturation J are smaller than in Fe, but much larger than in Py (cf. Chapter 4.2).

Wire dimensions: Length 70 nm, diameters 10 nm
Wire orientation: 4 perpendicular wires, crossed at the ends
Material: Cobalt (Co)
Exchange constant A: 1.3 x 10^{-11} J/m
Magn. polarization at saturation J:1.76 T
Gilbert damping constant α: 0.1
Tetrahedal mesh dimension: ≤ 3 nm
Field range: - 600 kA/m … + 600 kA/m (in x-y-plane)
Field sweeping speed: 10 kA/(m ns)

Fig. 4.2.31 shows the simulated system (left panel) and the resulting coercive fields for angles between 0° and 180° (right panel). Additionally, two or three further reversal positions, respectively, are depicted which can be recognized in the respective hysteresis loops.

The coercive fields have similar values as the results of the Fe simulations on similar systems (cf. Fig. 4.3.6, system with wires crossed at the ends); however, the angle dependence has a qualitatively different form here, with broad maxima around $(45+n \cdot 90)°$ instead of the sharp peaks in the Fe simulations.

Additionally, two or three further reversal positions occur which have not been seen in the Fe simulations. These additional reversal mechanisms are depicted more in detail in Fig. 4.3.24.

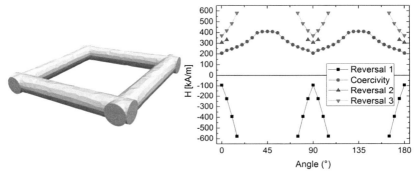

Fig. 4.3.23: 4-wire Co sample (left panel), simulated coercivities H_C and two to three additional reversal fields of this magnetic system (right panel). The angle 0° corresponds to a field orientation parallel to one of the wires.

For a sample orientation of 0° (Fig. 4.3.24, upper left panel), the first magnetization reversal step from positive saturation into a horseshoe state occurs before the external field vanishes. Such a behavior has been seen before, e.g., in the square sample with cylindrical cut A2 (cf. Fig. 4.2.25) or in all half-balls without cylindrical top-cut (cf. Figs. 4.2.17 and 4.2.18); however, is has not occurred in the four-fold Fe samples. The next steps lead over a vortex state (Fig. 4.3.25, 3rd panel in top row) and a second horseshoe state (4th panel in top row) to negative saturation.

If the sample is slightly rotated to 5°, the first magnetization reversal step (Fig. 4.3.24, upper right panel) results in an onion state, as depicted in Fig. 4.3.25 (2nd row, 2nd panel). The next steps lead to two further onion states, followed by negative saturation.

For a sample orientation of 10°, the second onion state vanishes. Taking into account the snapshot of the magnetization reversal process between onion 1 and onion 3 in Fig. 4.3.25 (3rd row, 3rd panel), a new non-diagonal onion state can be recognized; however, typical for domain wall processes, no additional step in the hysteresis loop occurs for this state.

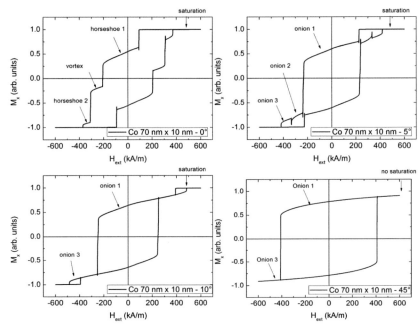

Fig. 4.3.24: Simulated hysteresis curves for the Co sample depicted in Fig. 4.3.23, exemplarily shown for orientations relative to the external magnetic field of 0° (i.e. parallel to one pair of wires), 5°, 10°, and 45°.

Finally, for the 45° sample orientation saturation is not completely reached within the simulated field range. However, since there are no energy barriers between the onion states at large positive and negative fields and the complete saturation, the magnetization reversal process is not influenced by this finding. The first onion state is extended to an "extreme onion" (bottom row, 2nd panel of Fig. 4.3.25) as it has been found before in the four-fold Fe sample with the wires crossed near the ends (Fig. 4.3.9). Magnetization reversal into another onion state happens via domain wall processes (3rd panel in 4th row of Fig. 4.3.25).

Concluding, it can be stated that changing the material parameters while the system geometry stays identical can lead to completely different hysteresis loops and reversal modes. Thus, especially for finding magnetic nano-systems which can be used as quaternary or even higher-order storage devices, it is very sensible to test several materials for the planned systems. Apparently, new materials may lead to novel and unexpected behavior – and can thus result in new functionalities.

Fig. 4.3.25: Magnetization reversal of the Co wire system from positive to negative saturation for sample angles of 0° (top row), 5° (2nd row), 10° (3rd row), and 45° (bottom row). The colors depict a magnetization orientation in positive (red) or negative (blue) x-direction; the external field direction is marked in each row.

4.4.   2-fold wire systems

While four-fold wire systems are of special interest for possible novel magnetic storage systems, the first attempts to produce magnetic micro-systems based on textile methods produced the best results for wrapped fine wires. Thus, this chapter examines the coercive fields of parallel magnetic wires, with the focus on irregularities, like different couplings between neighboring wires or shorter wires, i.e. broken wire parts. Finally, to allow for better comparison with the following chapters, a system with two parallel wires and additional balls at the ends is examined.

4.4.a. Angular dependence of the coercive fields – parallel wires
Wire dimensions: Lengths 70 nm / 70/3 nm, diameters 10 nm
Wire orientation: 2 wires, different orientations and distances
Material: Iron (Fe)
Exchange constant A: 2 x 10^{-11} J/m
Magn. polarization at saturation J: 2.1 T
Gilbert damping constant α: 0.1
Tetrahedal mesh dimension: ≤ 3 nm
Field range: - 800 kA/m … + 800 kA/m (field in x-y-plane)
Field sweeping speed: 10 kA/(m ns)

In order to examine the influence of neighboring magnetic wires on the coercivity without taking into account the crossing regions, this chapter concentrates on parallel pairs of wires in different distances. For comparison with the last chapter, one pair of perpendicular wires is added.

Fig. 4.4.1 shows the simulated geometries. In the coupled case (blue), two wires are oriented parallel, with an additional connector between them to enhance coupling. The same geometry without the connector, i.e. without a direct contact between both wires, is referred to as uncoupled (magenta). In the 1:3 coupled sample (green), two wires with an additional connector are simulated; however, one of the wires has only 1/3 of the usual length (70 nm). In the 1:3 shifted case (red), the shorter wire is shifted so that it ends in the middle of the other one. Finally, in the last sample, two wires are crossed perpendicularly at an asymmetric point (grey).

Fig. 4.4.1: Magnetic systems composed of two nano-wires, arranged in parallel / perpendicular pairs, with different coupling: parallel coupled (blue), parallel non-coupled (magenta), parallel 1:3 coupled (green), parallel 1:3 shifted (red), and perpendicularly crossed asymmetrically (grey).

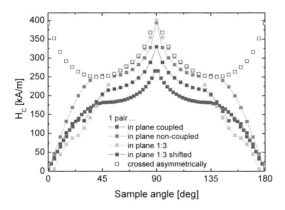

Fig. 4.4.2: Simulated coercivities H_C of the magnetic systems depicted in Fig. 4.4.1. The colors correspond to the sample colors in Fig. 4.4.1.

Fig. 4.4.2 shows the angular-dependent coercivities, simulated for the samples described above. Comparing the coercive fields of the crossed sample with Fig. 4.3.6, the coercive fields are higher for the 1x1 sample than in all 2x2 samples. Interestingly, the hard/easy axes are identical with those of the uncoupled samples and opposite to the behavior of the 2x2 samples in which the wires interpenetrate.

The values of the crossed sample are an upper boundary for all other samples. While the coercivities are very similar for the uncoupled sample within the field region 45° … 135°, the 1:3 coupled sample shows even slightly higher values around 90°, but significantly smaller coercivities outside the field region 60° … 120°.

The sample with in-plane coupling shows the lowest coercive fields. This can be attributed to the largest connected sample area and thus the smallest form anisotropy counteracting a rotation of the magnetization. The 1:3 shifted sample has lower coercivities around 90° than the 1:3 coupled one, since the more centered position of the short wire influences a larger area of the long wire.

Comparing the qualitatively and quantitatively differing angular dependencies of the coercive fields in Fig. 4.4.2, it becomes obvious that even small changes in the design of a nano-patterned structure can strongly influence the magnetization reversal of such systems.

In Fig. 4.4.3, hysteresis loops for the samples described before are shown. For the first two samples (coupled and non-coupled), the hard axis loops (0°) are obviously completely reversible, as has been recognized before for field sweeps along the symmetry axis (z-axis) of half-balls with a cylindrical hole in the middle (Fig. 4.2.3). In these samples, the easy axis loops (90°) show a nearly instantaneous switching of the magnetization.

For both 1:3 samples, a middle sample orientation (e.g. 45°) leads to a step in the hysteresis loop, as has been mentioned before in four-fold samples. It should be underlined that this behavior can apparently also appear in two-fold samples with a symmetry break. Additionally, it should be mentioned that also the easy-axis loop of the 1:3 coupled sample shows a broad, flat step on both sides of the loop.

Interestingly, extrapolating the steps for this loop and for both hysteresis curves at 45° back to $H_{ext} = 0$ results in positive magnetization values, opposite to the finding in four-fold samples, where an extrapolation to $M_x = 0$ for $H_{ext} = 0$ suggests a vortex state, and for $M_x < 0$ at $H_{ext} = 0$, an onion state can be expected.

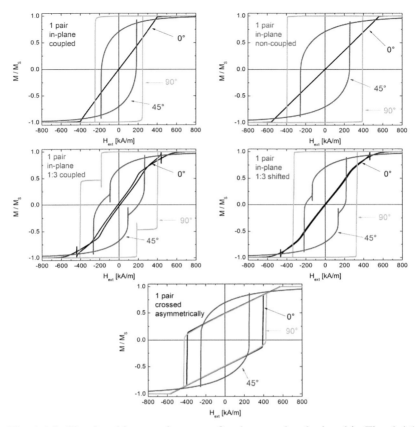

Fig. 4.4.3: Simulated hysteresis curves for the samples depicted in Fig. 4.4.1, exemplarily shown for orientations relative to the external magnetic field of 0° (i.e. parallel to one pair of wires), 45° and 90°. The colors in the panel descriptions correspond to the graphs colors in Fig. 4.4.2.

To examine this finding in detail and to understand magnetization reversal processes in the samples depicted in Fig. 4.4.1, snapshots of the magnetization during the reversal process for sample orientations of 0°, 45°, and 90° are shown in Figs. 4.4.4-4.4.7.

In Fig. 4.4.4, magnetization snapshots of the coupled sample are shown. In the top row, the reversal from positive to negative saturation is depicted for a sample angle of 0°. In this case, the reversal takes place via a vortex-like state with decreased stray fields in comparison to a process, in which the magnetization would relax for small fields (2nd panel in top row) due to the form anisotropy in identical directions in both wires.

For a sample angle of 45° (here the field vector is rotated, to avoid confusion due to the coloring with respect to M_x), oppositely, a field reduction from positive saturation leads to the magnetization in both wires being oriented parallel, before a decoupling of the magnetization in the wire ends leads to a wave-like magnetization (3rd panel) and finally to negative saturation.

In 90° sample orientation, finally, a perpendicular domain wall is built (2nd panel) which is moved from "top" to "bottom" of the sample, leading to a complete magnetization reversal after reaching the "lower" end of the pair of wires.

Fig. 4.4.5 shows the magnetization reversal of the pair non-coupled in-plane for the same angles. For 0°, the reversal works in the same ways as in the coupled sample described before.

For 45°, the process starts similarly; however, a domain wall is built in one of the wires (3rd panel) before the magnetization reversal process is finished.

The reversal for a sample orientation of 90° does not work with one common domain wall in both wires, as in the coupled sample, but with two separate domain walls in both wires.

In Fig. 4.4.6, the magnetization reversal of the 1:3 coupled pair from positive to negative saturation is shown. For 0°, the process is similar to those described for the other two samples. Here it can be seen that the "single" part of the longer wire reacts much earlier to the form anisotropy than the coupled part, with the magnetization in the latter staying longer oriented to the positive external field (2nd panel).

A similar effect can be recognized for 45°: After the usual alignment of the magnetization along positive y-axis due to the form anisotropy (2nd panel), the magnetization reversal starts in the short wire, i.e. in the coupled part (3rd panel), where the form anisotropy is reduced due to the coupling with the long wire and due to the reduced length of the short wire. This situation in which only the magnetization in the short wire is reversed is correlated with the step in Fig. 4.4.3.

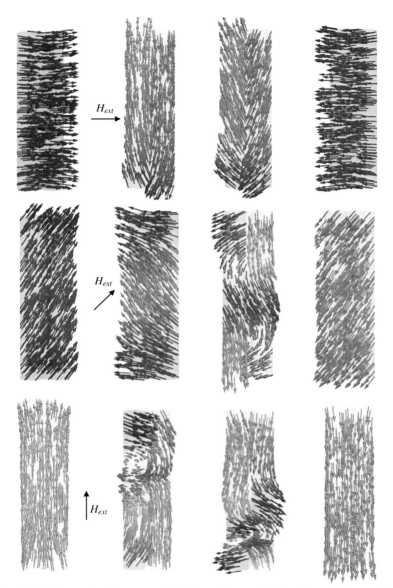

Fig. 4.4.4: Magnetization reversal of the pair **coupled in-plane** from positive to negative saturation for sample angles of 0° (top row), 44° (middle row), and 90° (bottom row). The colors depict a magnetization orientation in positive (red) or negative (blue) x-direction.

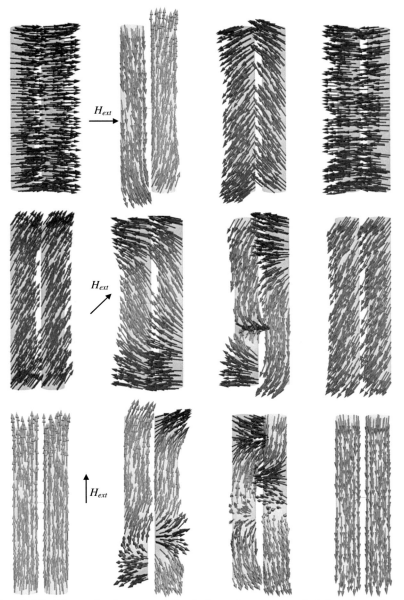

Fig. 4.4.5: Magnetization reversal of the pair **non-coupled in-plane** from positive to negative saturation for 0° (top row), 45° (middle row), and 90° (bottom row).

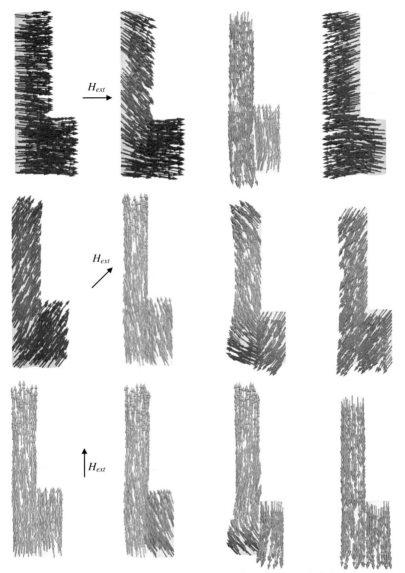

Fig. 4.4.6: Magnetization reversal of the **1:3 pair coupled in-plane** from positive to negative saturation for 0° (top row), 45° (middle row), and 90° (bottom row). The colors depict a magnetization orientation in positive (red) or negative (blue) x-direction; the external field direction is marked in each row.

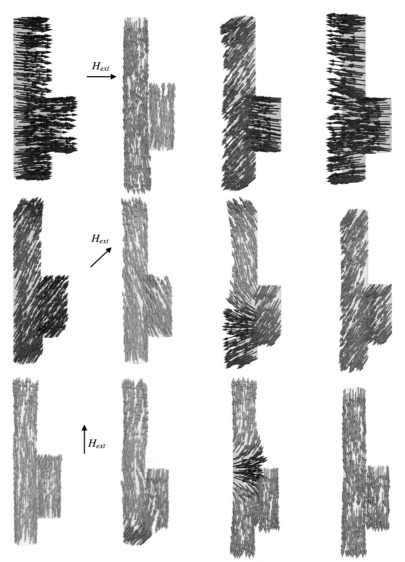

Fig. 4.4.7: Magnetization reversal of the **1:3 shifted pair coupled in-plane** from positive to negative saturation for 0° (top row), 45° (middle row), and 90° (bottom row). The colors depict a magnetization orientation in positive (red) or negative (blue) x-direction; the external field direction is marked in each row.

This effect is even stronger for 90°, where the short wire reverses completely before the long one starts the reversal process (3rd panel). This situation is the explanation for the broad step in the hysteresis loop which could be recognized in Fig. 4.4.3.

It should be mentioned that the finding of a step with $M_x > 0$ in the hysteresis loops can be attributed to a small part of the magnetization (here: ~ ¼ of the overall magnetization) being already reversed while the major part is still oriented oppositely.

Fig. 4.4.7 shows the magnetization reversal of the 1:3 shifted pair of wires. For 0°, similar to the situation of the coupled 1:3 pair, the short wire stays oriented along positive saturation for longer and is reversed to negative saturation by smaller negative fields (3rd panel).

For 45°, magnetization reversal again starts first in the short wire, leading to the step in Fig. 4.4.3.

However, at a sample orientation of 90° to the external field, the reversal of the short wire (2nd panel) does not result in a stable intermediate state, i.e. a step in the hysteresis loop, but triggers the magnetization reversal in the long wire via a domain wall process (3rd panel). Both processes are nearly indistinguishable in Fig. 4.4.3.

In Fig. 4.4.8, the magnetization reversal processes of the asymmetrically crossed sample are depicted. For a sample orientation of 0°, firstly the magnetization in the "vertical" wire is directed parallel to this wire due to the form anisotropy, when the external field is decreased (2nd panel). The coercive field in Fig. 4.4.3 is defined by the switch of the magnetization in the "horizontal" wire, followed by the orientation of the magnetization in the "vertical" wire along negative saturation.

For 45°, the rotation of the magnetization in the "vertical" wire is accompanied by a magnetization rotation with opposite rotational direction in the "horizontal" wire. Finally, both processes are finalized by a further rotation into negative saturation which defines the coercive field.

Interestingly, the magnetization reversal for 90° includes a domain wall process (3rd panel), which has not been observed for the 0° orientation. Apparently, the asymmetric crossing position of the wires yields indeed a difference between 0° and 90°, which is also visible in the slightly different hysteresis loops in Fig. 4.4.3.

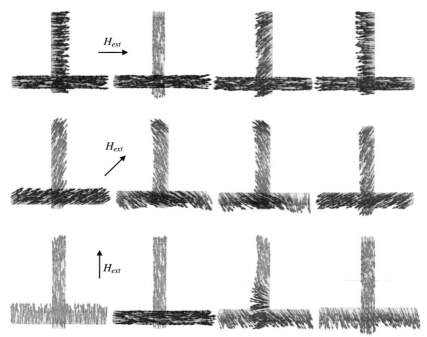

Fig. 4.4.8: Magnetization reversal of the pair **crossed asymmetrically** from positive to negative saturation for sample angles of 0° (top row), 45° (middle row), and 90° (bottom row). The colors depict a magnetization orientation in positive (red) or negative (blue) x-direction; the external field direction is marked in each row.

4.4.b. Comparison with wire sample with extended ends
Wire dimensions: Length 70 nm, diameter 10 nm
Wire orientation: 1 wire with half balls (diameter 20 nm) at ends
Material: Iron (Fe)
Exchange constant A: 2 x 10^{-11} J/m
Magn. polarization at saturation J: 2.1 T
Gilbert damping constant α: 0.1
Tetrahedal mesh dimension: ≤ 3 nm
Field range: - 600 kA/m … + 600 kA/m (field in x-y-plane)
Field sweeping speed: 10 kA/(m ns)

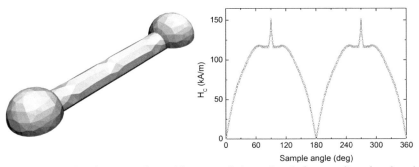

Fig. 4.4.9: 2-wire sample with extended ends (left panel), simulated coercivities H_C of this magnetic systems (right panel). The angle $0°$ corresponds to a field orientation perpendicular to the wires.

To allow for an easier comparison of the results of two-fold samples with other samples examined within this work, additionally a single wire with extended ends has been simulated. The half-ball construction depicted in Fig. 4.4.9 (left panel) has also been used in the simulations of the three-fold and six-fold samples (Chapters 4.5, 4.6).

As can be seen in Fig. 4.4.9 (right panel), the coercivities (simulated here for 360° in 1° steps) have a similar angle dependence as both pairs with 2 wires of identical length (cf. Fig. 4.4.2), but with significantly smaller absolute values of the coercive fields.

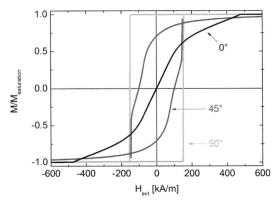

Fig. 4.4.10: Simulated hysteresis curves for the sample depicted in Fig. 4.4.9, exemplarily shown for orientations relative to the external magnetic field of $0°$ (i.e. perpendicular to the pair of wires), 45° and 90°.

In the hysteresis loops for 0° and 45°, however, a difference in the shape can be recognized, comparing Fig. 4.4.10 with Fig. 4.4.3: Here, the hysteresis loop seems to contain two different fractions, resulting in two different slopes in the 0° curve and a loop with an abrupt magnetization change for 45°, opposite to the quite "round" 45° loops in Fig. 4.4.3.

This difference can indeed be attributed to the existence of the additional half-balls at both ends of the single wire. As Fig. 4.4.11 shows, these balls offer a different magnetization reversal behavior as the wire itself, due to their completely different form anisotropy. For 0° (i.e. the magnetic field oriented perpendicular to the wire, top row), the magnetization reversal starts in the wire, due to the form anisotropy which favors the magnetization orientation parallel to the wire. When the external field vanishes (3rd panel), the form anisotropy of the wire is sufficient to align the magnetization in the whole sample parallel to the wire, leading to a nearly vanishing coercive field ($H_C = 0.1$ kA/m). Similarly, the reversal process is finished first in the half-balls (4th panel), while numerically larger negative fields are necessary to reach negative saturation in the wire.

A similar behavior can be found for a sample orientation of 45° (middle row). The magnetization in the wire rotates to the orientation parallel to this form, thus favoring one rotational direction for the magnetization in the half-balls. Only when the half-balls have nearly reached negative saturation (3rd panel), the magnetization in the wire follows this process (4th panel).

For 90°, the saturation magnetization direction of the wire is identical with the direction favored by the form anisotropy. Thus, the magnetization does not change orientation in this part of the system until numerically large negative fields are reached. Thus, both half-balls do not reverse commonly, since opposite to the 45° case, this process cannot be triggered by an initial rotation of the wire magnetization. The magnetization reversal inside the wire occurs after domain walls to both half-balls have been built.

These magnetization snapshots show clearly that the hysteresis loops of the system under examination do indeed consist of two components, related to the wire and the half-balls, respectively. However, although the half-balls influence the magnetization reversal in the wire, working as additional magnetic fields at both ends, the angle dependent coercive fields change only quantitatively, not qualitatively.

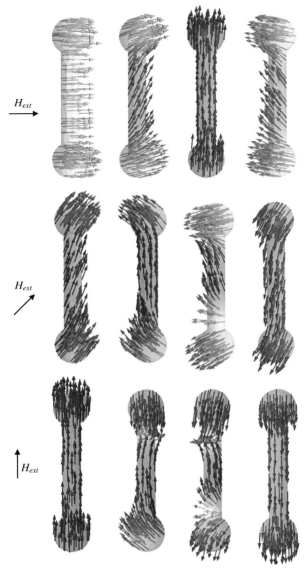

Fig. 4.4.11: Magnetization reversal of the parallel pair with extended ends from positive to negative saturation for sample angles of 0° (top row), 45° (middle row), and 90° (bottom row). The colors depict a magnetization orientation in positive (red) or negative (blue) **y-direction**; the external field direction is marked in each row.

4.5. 3-fold wire system
Wire dimensions: Lengths 70 nm, diameters 10 nm
Wire orientation: 3 wires with half balls (diameter 20 nm) at ends
Material: Iron (Fe)
Exchange constant A: 2 x 10^{-11} J/m
Magn. polarization at saturation J: 2.1 T
Gilbert damping constant α: 0.1
Tetrahedal mesh dimension: ≤ 3 nm
Field range: - 600 kA/m … + 600 kA/m (field in x-y-plane)
Field sweeping speed: 10 kA/(m ns)

A threefold sample has been simulated for comparison with the MOKE experiments on Co/CoO(111) (Chapter 3.4). The ends of the three wires have been connected by half-balls, as shown in Fig. 4.5.1 (left panel).

While former theoretical symmetry considerations and the evaluation of the experimental results of the Co/CoO(111) sample imply a six-fold symmetry of the coercive field, the simulations give rise to a different graph (Fig. 4.5.1, right panel): The coercive field (black dots) does indeed show a slightly six-fold symmetry, however, the variation of this value is very small, quite opposite to simulations of four-fold or two-fold systems. Significant variations of the hysteresis loops (Fig. 4.5.2) occur not by changes in the coercivity, but by large variations of the one or two additional reversal fields.

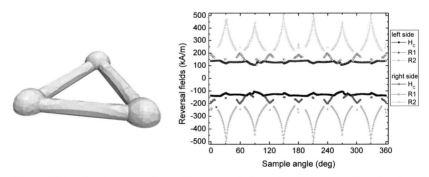

Fig. 4.5.1: 3-wire sample with extended ends (left panel), simulated coercivities H_C and two additional reversal fields of this magnetic system (right panel). The angle 0° corresponds to a field orientation parallel to one of the wires.

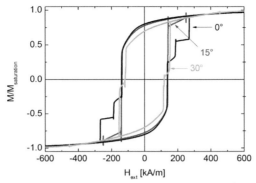

Fig. 4.5.2: Simulated hysteresis curves for the sample depicted in Fig. 4.5.1, exemplarily shown for orientations relative to the external magnetic field of 0° (i.e. perpendicular to the pair of wires), 15° and 30°.

As Fig. 4.5.2 shows, the hysteresis loop for 0° depicts two additional reversal fields, causing two steps with different slopes. The first step in the field area of ~ -130 kA/m to -190 kA/m can be extrapolated to the origin of the coordinate system. Such a behavior is typical for vortex states. The second step in the area of ~ -190 kA/m to -280 kA/m has a clearly different slope.

Fig. 4.5.3 shows that the magnetization reversal for 0° (top row) occurs indeed via an onion state (2nd panel), which is correlated with the "round" slope of the hysteresis loop around 0 kA/m, followed by a vortex state (3rd panel) and a state with two domain walls, in which only the upper half-ball is not yet reversed (4th panel). The latter is connected with the second step.

The hysteresis curve for a sample orientation of 15° shows only one step, which is again different from both 0° steps. Especially, this step is approaching the reversible part of the hysteresis loop, near negative saturation, without a "jump" in the magnetization. As can be seen in Fig. 4.4.14 (middle row), this step can be identified as a second onion state (3rd panel) following the first onion state (2nd panel) which could also be seen in the reversal for 0° sample orientation.

For a sample orientation of 30°, the additional step in the hysteresis loop can be attributed to an additional domain wall process (not shown here) between both onion states (2nd and 3rd panel). The broad, flat step ending at ~ -460 kA/m, finally, results from the rotation of the magnetization in the wire perpendicular to the external field (compare 3rd and 4th panel).

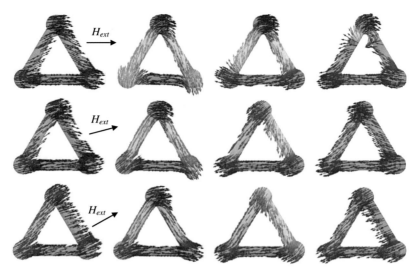

Fig. 4.5.3: Magnetization reversal of the parallel pair with extended ends from positive to negative saturation for sample angles of 0° (top row), 15° (middle row), and 30° (bottom row). The colors depict a magnetization orientation in positive (red) or negative (blue) y-direction; the external field direction is marked in each row.

Comparing the magnetization snapshots of the three sample orientations in Fig. 4.5.3 and the three hysteresis loops in Fig. 4.5.2, it becomes clear why the coercive fields do not vary significantly: The main reversal processes are identical for all orientations besides 0°. Although additional reversal processes can occur for different sample orientations (see angle-dependent reversal fields in Fig. 4.5.1), these reversal mechanisms occur in most cases only after reaching coercive fields. Additionally, the vortex state for 0° is taken at nearly the same external magnetic field as the second onion state for angles up to ~ 20°. The decrease of the coercivity around 30° is correlated with the existence of a broad, flat "step" (green dots in Fig. 4.5.1).

Slight asymmetries in the hysteresis loops, which can be recognized by comparing the positive and the negative values in Fig. 4.5.1 (right panel), are due to small irregularities in the simulated grid. In a real sample, such irregularities can also be expected. Averaging over several single three-fold systems within a real sample, it can be expected that experimental results will produce less pronounced steps, where these asymmetries occur in the simulation.

4.6. 6-fold wire system

Since 6-fold wire samples can exhibit more than 2 stable states for a vanishing external field [Bla13a], similar to the 4-fold samples examined before, six 6-fold samples with different corner connections have been simulated (Fig. 4.6.1).

Wire dimensions: Lengths 60-80 nm, diameters 10 nm (20 nm)
Wire orientation: Six-fold, partly half-balls / cylinders at ends
Material: Iron (Fe)
Exchange constant A: 2 x 10^{-11} J/m
Magn. polarization at saturation J: 2.1 T
Gilbert damping constant α: 0.1
Tetrahedal mesh dimension: ≤ 3 nm
Field range: - 500 kA/m … + 500 kA/m (field in x-y-plane)
Field sweeping speed: 10 kA/(m ns)

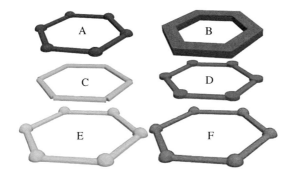

Fig. 4.6.1 gives an overview of the system geometries. The half-ball or cylinder diameters, respectively, are 20 nm. The wire diameters are 20 nm (sample B) or 10 nm (else). The maximal lateral dimensions, measured between the middle of the half-balls / cylinders at opposite corners, is 120 nm (sample B), 140 nm (samples A, C, and D), or 160 nm (samples E and F). In sample E, the half-ball base planes are in the same plane as the wire base planes; in sample F, the half-balls are up-shifted. For a sample orientation of 0°, the external magnetic field is parallel to one of the wires.

In Fig. 4.6.2, the angle-dependent coercive fields are depicted for the samples shown in Fig. 4.6.1. The colors of the plots correspond with those of the samples above.

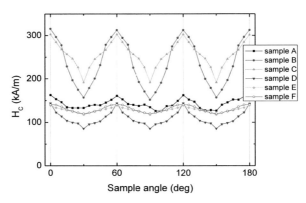

Fig. 4.6.2: Simulated coercivities H_C of the magnetic systems depicted in Fig. 4.6.1. The colors correspond to the sample colors in Fig. 4.6.1.

Samples B and C – the only samples without additional corner connections – show much larger coercive fields and stronger anisotropies than the other samples. As could be seen before in samples of other geometry, the extended corner connections work as a starting point for the reversal process, due to the reduced form anisotropy in these areas. In the same way, they support the magnetization reversal here, too.

Sample A shows significant deviations from a pure six-fold anisotropy. This behavior can be understood by taking into account Fig. 4.6.3, where the hysteresis loop for a sample orientation of 30° (upper left panel) shows at least four steps during magnetization reversal. The magnetization reversal processes in each wire and each half-ball are not very stable against small modifications of the geometry of the respective sample part. Thus, even very small deviations in the meshing result in macroscopic changes of the hysteresis loop in this sample.

One or two steps during magnetization reversal can also be recognized for the other samples with corner connections around the 30° orientation. However, magnetization reversal in these samples is apparently more stable against small deviations of a perfect geometry.

Up-shifting the half-balls in sample F changes the coercive fields only marginally, compared to sample E, while exchanging the half-balls of sample A by cylinders (sample D) reduces the coercivities as well as the number of steps in the slope of hysteresis loops significantly.

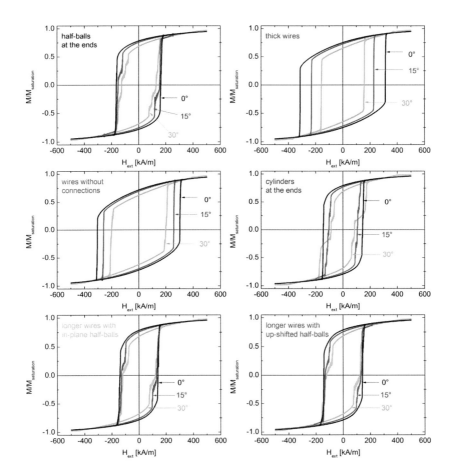

Fig. 4.6.3: Simulated hysteresis curves for the samples depicted in Fig. 4.6.1, exemplarily shown for orientations relative to the external magnetic field of 0° (i.e. parallel to one pair of wires), 15° and 30°. The colors in the panel descriptions correspond to the graphs colors in Fig. 4.6.2.

The reversal process is depicted by snapshots of the magnetization in Fig. 4.6.4 for a sample orientation of 30°, which shows the strongest differences in the hysteresis loops.

The first step of the magnetization reversal, starting from positive saturation, is always an onion state, independent of the exact sample geometry (2nd panel in each row). The further process can proceed either via additional stable states in which complete wires have already reversed magnetization (e.g. 3rd panel for samples A, D, E, and F), resulting in steps in the hysteresis loops, or via domain wall processes (e.g. 3rd panel for sample B).

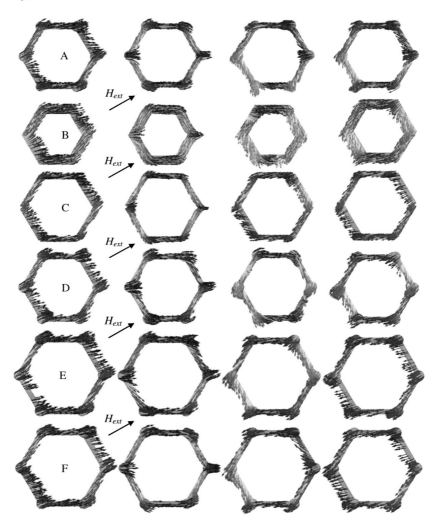

Fig. 4.6.4: Magnetization reversal for a sample orientation of 30° from positive to negative saturation for samples A-F. The colors depict a magnetization orientation in positive (red) or negative (blue) y-direction; the external field direction is marked in each row.

Concluding, magnetization reversal processes in all six-fold wire systems follow a similar scheme; however, dependent on the exact geometry, the number of steps in the hysteresis loop and the reaction of the system on slight asymmetries in the simulated meshes can vary strongly.

5. Outlook

In this chapter, an overview will be given of the planned next step related to the research topic of bit-patterned media with more than 1 bit per magnetic pattern. Starting with results of MathCad simulations for novel anisotropies, future experiments as well as the next simulations on more enhanced structures are described, followed by a collection of ideas in which other systems the novel magnetic structures can be utilized. Additionally, an overview of the technical parameters related to real storage devices using the new magnetic structures is given.

5.1. Systems with novel anisotropies

This sub-chapter aims at finding novel mathematical terms for the anisotropies to describe the experimental findings and micromagnetic simulations which could not yet be fitted with the usual phenomenological descriptions.

Firstly, several possibilities to create sixfold anisotropies have been tested. Fig. 5.1.1 shows the results of different models. While the black dots depict, for comparison, the usual term $\cos(6\phi)$, the green dots results from a threefold anisotropy $\cos(3\phi)$. The latter indicates clearly that a threefold anisotropy leads to a threefold symmetry of the coercive fields, which has not been seen in the experiments nor in the results of the micromagnetic simulations.

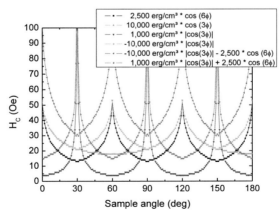

Fig. 5.1.1: Coercive fields for calculations of different threefold (green dots) and sixfold (black, dark blue, and cyan dots) anisotropies, complemented by combinations of different anisotropy formulae.

Next, two absolute value functions (dark blue and cyan dots) have been tested, since such a function has been shown to fit well for special exchange bias systems before [Til06a]. However, the shapes of the angle dependence of the coercivity found in the micromagnetic simulations – with sharp maxima and minima (Fig. 4.6.2) or with broad plateau-like maxima and dips between (Fig. 4.5.1) – cannot be reproduced by these functions. For the combinations of absolute value and cosine functions, the results look similar.

While the cyan, pink, and the dark yellow function in Fig. 5.1.1 exhibit indeed sharp maxima and minima, the forms of the different sixfold micromagnetic simulations (Fig. 4.6.2) is not perfectly reproduced.

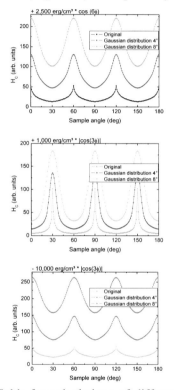

Fig. 5.1.2: Coercive fields for calculations of different sixfold anisotropies, smoothed by assuming Gaussian distributions of the anisotropy fields.

As a next step, the three single sixfold anisotropies from Fig. 5.1.1 have been modeled taking into account a Gaussian distribution of the anisotropy axes, with different full widths at half maximum of 4° and 8°, respectively. The results are depicted in Fig. 5.1.2.

Obviously, this procedure smoothes the sharp maxima and reduces the broad minima (cf. especially the middle panel). However, comparing with most of the sixfold anisotropies found in the micromagnetic simulations (Fig. 4.6.2), the Gauss distribution of the anisotropies does not help to reproduce these results. Since the form of the micromagnetic sixfold coercivities reminds at twofold simulations (Fig. 4.4.2), in a next step, combinations of three twofold anisotropies, shifted for 60° to each other, are examined.

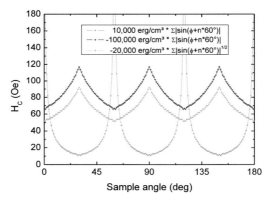

Fig. 5.1.3: Coercive fields calculated for different combinations of 3 x twofold anisotropies (see inset for formula).

As Fig. 5.1.3 shows, it is possible to gain more pronounced sharp minima in this way. However, even introducing more distinct energy minima by using the square root function (cyan dots) does not lead to the inflection points which have been recognized in some of the micromagnetic simulations of the sixfold samples (Fig. 4.6.2).

Possible reasons for this difference are, e.g., anisotropies which have not been taken into account here, like a rotational anisotropy. What is more likely, however, is a change in the anisotropy formula for different angles. It can be assumed that especially the interaction between the wires and the (half-)balls at their ends may lead to completely new physical properties, such as angular-dependent energy formula, similar to exchange bias systems with varied field cooling directions.

One possibility to find out more about the reasons for this difference is the examination of the simulated samples by time-dependent micromagnetic simulations. As in FMR (FerroMagnetic Resonance) or BLS (Brillouin Light Scattering), applying a short magnetic field pulse in saturation should result in a precession of the simulated magnetization. Future simulations can show whether these principles can be transferred to the nano-structured samples simulated here at all. If so, these time-resolved simulations will help understanding the anisotropies of the threefold and sixfold wire samples.

An experimental result which could not be reproduced by the "usual" anisotropies has been found in the system Co/CoO(110) (Fig. 3.6). While the calculations of the angle dependent coercivities, using typical uniaxial and fourfold anisotropies, resulted in a step between neighboring maxima, such a step could not be recognized in the measurement results.

Fig. 5.1.4 shows the coercivities calculated with the typical fourfold anisotropy term $\sin^2(\phi) \cos^2(\phi)$ and a uniaxial term modeled as $\pm|\sin(\phi)|$. While the negative sign before the absolute value term results in a splitting of the maxima around 0° and 180° (black dots), which is not desired, the twofold terms with positive sign can indeed generate the desired difference between higher and lower maxima without the problematic steps near the minima. However, the correlation between the coercivities in the maxima and the minima does not correlate completely with the experimental findings. Thus, as a next step, reducing the maxima height by using a Gaussian distribution is examined.

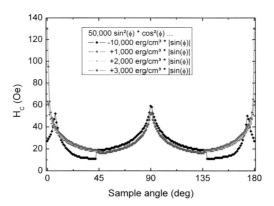

Fig. 5.1.4: Coercive fields calculated for different combinations of twofold plus fourfold anisotropies (see inset for formula).

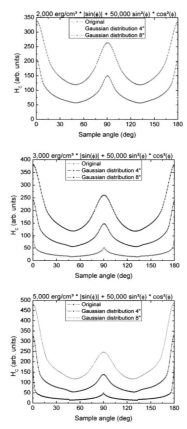

Fig. 5.1.5: Coercive fields calculated for different combinations of twofold plus fourfold anisotropies, smoothed by Gaussian distributions.

Fig. 5.1.5 shows Gaussian distributions of different combinations of uniaxial and fourfold anisotropies. While the ratio of the larger to the smaller maximum changes for different anisotropy combinations and Gaussian distributions with different full widths at half maximum, the ratio of the smaller maximum to the minima does not change significantly. Thus, by applying a Gaussian distribution, the experimental results with a quite narrow lower maximum (Fig. 3.6) cannot be reproduced.

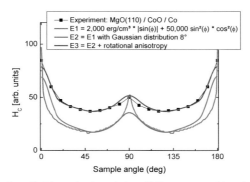

Fig. 5.1.6: Coercive fields calculated for different combinations of twofold plus fourfold anisotropies, partly smoothed by Gaussian distributions and complemented by a rotational anisotropy.

Fig. 5.1.6 shows further tests to reproduce the experimental values (black dots). As described before, neither the pure combination of an absolute value term as uniaxial anisotropy and a typical fourfold anisotropy (orange line) nor the same equation with a Gaussian distribution of the anisotropies (green line) can reproduce the complete angular dependence found in the experiment. However, it is possible to fit the measured results very well if an additional up-shift of the coercivities is allowed (red line).

Such an up-shift could be explained by a rotational anisotropy which has been introduced by Stiles and McMichael [McM98, Sti99] for polycrystalline exchange biased samples. They found only some of the grains (in the antiferromagnet) to be sensitive to the unidirectional field, leading to an isotropic shift of the spin wave frequencies in FMR measurements. It can be assumed that similarly to this effect, such a rotational anisotropy would also cause an isotropic up-shift of the coercivities, if not all parts of the sample are equally sensitive to the external field. This idea is supported by former BLS measurements [Bla07a] which have proven the existence of a rotational anisotropy in a sample with identical construction as the one examined here.

While the coercivities measured experimentally and simulated for both uncoupled systems could be reproduced well by the usual fourfold anisotropy term (cf. Fig. 3.13), the coupled wire systems showed different shapes of the angle-dependent coercivities. In fig. 5.1.7, coercive fields are depicted for several further possibilities to define fourfold anisotropies. Opposite to the sixfold anisotropies, the superposition of some twofold anisotropies does here not support the formation of relatively sharp minima, but leads to even broader minima. The best fits to the simulated wire system crossed at the ends is indeed given by the original function $\sin^2\phi \cos^2\phi$.

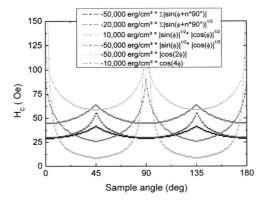

Fig. 5.1.7: Coercive fields calculated for different fourfold anisotropies (see inset for formula).

However, the coercive fields of the coupled crossed system, exhibiting small additional peaks along the hard axes, can be explained taking into account a former finding which has been published in [Til09]: Since the form anisotropy strongly suppresses a magnetization orientation perpendicular to the wires, the magnetization switches more probably via 180°, i.e. directly from one easy direction to the other. In such a way, an additional apparently eightfold anisotropy can occur, as it can be recognized in the coupled crossed system.

5.2. Proposals for technological solution

To get nano-structures samples similar to the best simulated systems, an application has been written in the biannual call of the Karlsruhe Nano Micro Facility (KNMF), Karlsruhe, Germany, which has been successfully evaluated.

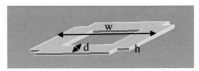

Fig. 5.2.1: CAD-depiction of the structure element produced in the KNMF in near future, with the element width w, the wall diameter d, and the height h.

The first set of samples, structured like in Fig. 5.2.1, is being produced in the moment in the KNMF. It is planned to develop systems with the following parameters:

w = 400 nm, d = 100 nm, h = 10 nm
w = 200 nm, d = 50 nm, h = 10 nm
w = 160 nm, d = 40 nm, h = 10 nm
w = 80 nm, d = 20 nm, h = 10 nm

Identical sample heights allow for metalizing all sets of samples on the same waver at the same time. The dimensions have been chosen according to discussions with the e-beam specialists from Karlsruhe, to get accurate samples similar to our previous simulations. Future applications for further samples will be based on the experience gained in the recent project.

Since the simulations deal with Fe samples, iron has also been chosen as magnetic material which will be introduced in the lithographed areas. An additional cap layer from titan will be used to prevent oxidation of the Fe.

The structured areas will be 1 μm x 1 μm, with the single objects arranged as depicted in Fig. 5.2.2.

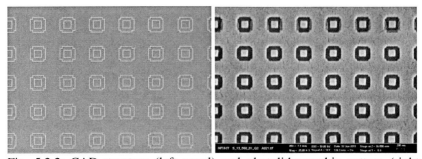

Fig. 5.2.2: CAD-structure (left panel) and photolithographic pattern (right panel) corresponding to the structure simulated in Fig. 4.3.19 (lower right panel). SEM picture taken in the KNMF, Karlsruhe.

These four samples will be characterized by SEM firstly and afterwards measured by MOKE and D-MOKE. The results will be compared with the simulations and be used to fit the theoretical results to the experiment. After this process, the next applications at the KNMF will be focused on smaller structures, different materials (e.g. Co due to the interesting results of the simulations of Co structures), or more sophisticated structures.

The next simulation will, on the one hand, concentrate on more sophisticated structures, in order to examine the possibility to store even 3 bits of data in one magnetic particle, correlated with 8 stable states.

On the other hand, the possibilities of ultrafast magnetic field pulses in magnetic nano-dots triggering the vortex core reversal in these systems shall be examined further. While this thesis concentrates on the idea of more than one bit per magnetic feature, possibilities of faster switching in magnetic dots are still of large technological interest, especially in combination with the bit-patterned media consisting of dots, recently developed by Toshiba [Tos10].

Additionally, further materials will be tested in the simulations, to find possible other interesting features and to get an overview which materials might be used in real hard disk platters. This is an important aspect for the transfer of basis research, as performed here, into a future standard technology.

Besides, 3D structures are planned to be simulated, according to the following challenges:

Is it possible to combine 2 layers of quaternary storage particles without losing the additional stable states at remanence?

Can in-plane and out-of-plane magnetization measurements be combined to measure two or more independent features of the magnetic particles? Does it make a difference if the single particles are grown on a stepped substrate, allowing for a growth under a defined angle to the base plane? Can such a tilted growth help to distinguish between the magnetization in different bars, especially in triangular particles?

And what about a stack of squares with different dimensions, switching at different external magnetic fields? Such a system might be able to combine the advantages of the Racetrack memory and the bit-patterned media being developed by Toshiba.

Besides the more sophisticated structures, 3D structures appear most interesting to be examined in the near future. They may be a further step on the way to overcome the limits of Moore's law.

Additional to these samples with pure ferromagnetic particles, more complex systems can be structured and experimentally examined. A test environment for current-induced switching can be structured, allowing for measurements of magnetization reversal triggered by current pulses – either with "simple" magnetic elements, like in former ferrite core memories, or even as MRAMs (Fig. 5.2.3). MRAMS and other complete spin valve elements could be structured according to the quaternary structures described in this thesis (Fig. 5.2.4). In such a way, a bridge can be built from the theoretical examination of single magnetic particles to the final goal of introducing such particles into bit-patterned media, MRAMs or other spintronic elements.

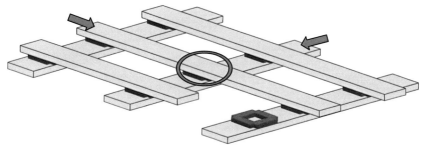

Fig. 5.2.3: Principle of MRAM particles with word and bit lines. A current pulse through both top and bottom lines is necessary to switch one selected position; the "half-selected" positions (experiencing only a current pulse through one of the lines) are not influenced. The dimensions of the "crossing points" as well as the distances between them can recently be reduced to about 80 nm x 80 nm, according to first experiences with lithography in the KNMF, Karlsruhe.

With a magnetic cell dimension of 80 nm x 80 nm, the word and bit lines in an array should have similar dimensions of about 100 nm x 100 nm. This order of magnitude is lower than recent typical values of ~ 250 nm width of the word and bit lines [Tan10]. However, even bit patterned media with 40 nm bit dimensions have already been proven to allow for sub 10^{-4} on-track error rates [Gro10]. Distances between single bits in bit patterned media can be lowered to ~ 5 nm in samples used for research [Yan11], which is also a typical value for shingled magnetic recording [Ven12]. Similar distances have to be approached subsequently with the novel system.

For values of word and bit line width 250 nm, the write field efficiency has been calculated to reach values of the flux conversion efficiency H/I ~ 10 Oe/mA for 100 nm spacing between bit and word line and H/I ~ 4 Oe/mA for 400 nm spacing [Tan10]. External fields of ~ 2.5 kOe, a value which was found to be typical for the saturation field of the nano-objects simulated in this thesis, would thus be correlated with a switching current of minimally 250 mA. This value is about twice as high as usual in recent MTJ switching processes [Tan10]. It can be reduced, e.g., by varying the dimension of the magnetic particles, by changing the materials, or by reducing the spacing between bit and word line.

In hard disk drives, coercivities of 4.800 kOe are typical for recent thin film recording media; write heads with $Fe_{65}Co_{35}$ used today have flux densities of 2.45 T [Tag01, Zhu03]. For this purpose, the fields necessary for switching the novel quaternary nano-particles fit ideally into recent technological parameters.

Such an MRAM cell utilizing the novel nano-wire system developed in this thesis would be able to exhibit more than the conventional two different tunnel resistances: With the free layer built as quaternary wire system, it would show four different states; with both – free and fixed – layers built as quaternary wire systems, even up to eight states are possible, in dependence on the switching fields of both layers.

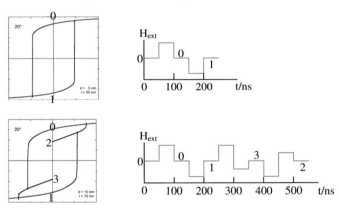

Fig. 5.2.4: Schematic of a common switching process with one bit (two states) per storage cell (upper panel) and of the switching process used for the novel storage cell with two bits (four states) (lower panel). Values for t have been chosen according to the simulation procedure shown in Fig. 4.3.2(a).

Opposite to typical switching processes, the switching field sequence has to be modified if the novel system is used. Fig. 5.2.4 shows a schematic of the common switching sequences and the novel sequences to reach the four different states. This scheme can be extended respectively for systems with more than four states at remanence.

6. Conclusion

Magnetic storage media are a topic of great interest for technological and fundamental research. Again and again, new materials with novel properties, advances in nano-structuring, or even completely new magnetization reversal mechanisms with unexpected properties are reported, such as switching the magnetic structure from AFM to FM in colossal magnetoresistive devices using short laser pulses, exceeding switching speeds from gigahertz to the terahertz region [Li13].

Due to the increasing problems of following Moore's law to higher data storage densities by decreasing the size of one bit, this thesis instead concentrates on the idea of enhancing the number of bits represented by one single magnetic particle.

To investigate such a possible new way of addressing bits in novel magnetic storage media, magnetic nano- and micro-structures have been examined in experiment and simulation, compared with calculations of magnetic anisotropies defining the magnetic properties of nano-particles.

Firstly, magnetization measurements by MOKE have underlined the consistency of some of the samples under investigation with common mathematical formulations of magnetic anisotropies, while other samples showed the necessity of finding novel anisotropy models to fit the experimental results. Measurements on a nano-dot structured sample showed the technical possibility to use the MOKE setup established in the Silesian University of Technology for further experiments on nano-structured samples with differently-shaped particles. Such a nano-structured sample, based on the results of the simulations shown in this thesis, is still being prepared in the KNMF, Karlsruhe, Germany. Trials to use textile-based structured samples, e.g. coated or screen-printed glass substrates, as model systems, however, have been stopped after having proven the impossibility to measure such samples by MOKE.

Simulations with Magpar have shown different magnetization reversal processes and related hysteresis loops as well as possible stable intermediate states. Such intermediate states – which are stable when the external magnetic field vanishes – result in more than the usual two states at remanence, offering the possibility of creating quaternary or even higher-order magnetic storage media.

In first simulations of magnetic half-balls, which are used in bit-patterned media and are very often investigated candidates for novel magnetic storage systems, such stable intermediate states could not be detected. Thus, fourfold and other magnetic wire systems were examined. Improving the systems under examination more and more, led to a nano-particle design which showed to be robust against scaling in a broad range of dimensions – a very supportive property in designing areas of smaller and smaller nano-particles, leading to increased data storage densities.

Further investigations of nano-particles with n-fold symmetry improved the understanding of the relations between magnetization reversal processes and sample symmetries, resulting in first approaches for higher-order systems, possibly allowing storage of 3 or even 4 bits per nano-particle.

In the outlook section, ideas for further simulations and experiments are described, starting from novel particle geometries to 3D structures and possible MRAMs including the nano-structure developed in this thesis.

In the future, collaboration with a hard disk drive producer is intended. The following technical parameters are likely to be obtained when the quaternary system simulated in this thesis is used in real hard disk drives:

According to first lithography results in KNMF, Karlsruhe, the sample structure suggested here (cf. Fig. 5.2.2) can be reduced to ~ 80 nm x 80 nm without loss of accuracy. Further dimensional decrease has to be examined in the future to understand which accuracy of the proposed structure is possible at which cell size, compared by the respective simulations of the disturbed geometries.

A cell size of 80 nm x 80 nm, with 2 bits per cell, results in a data density of 200 Gbit/in², which is still lower than recent HDD data densities of 750 Gb/in² on a two-sided hard disk platter [Fon13]. Using two sides as well and decreasing the dimensions of the cell to 60 nm x 60 nm would result in the same areal data density. Further enhancement of the data density requires examinations of scaling behavior in simulation and experiment.

The read/write times of hard disk drives are dominated by seek times in the order of about 10 ms and latency times of a few ms. In the simulations depicted in this thesis, a time interval of 50-100 ns is necessary for a complete magnetization reversal from positive to negative saturation; the switching process itself is performed within ~ 10 ns (without any trials to optimize these values by using, e.g., precessional switching). Thus, the read/write times of a potential future hard disk drive with the quaternary system suggested in this thesis are almost not influenced by the switching times of the single bits.

The coercive fields in the simulations of iron nano-wires in this thesis are ~ ± 200 kA/m or ~ ± 2.5 kOe, which would be the necessary magnetic fields to switch the single nano-particles on a hard disk platter.
Typical switching fields of recent hard disk materials are in a similar order of magnitude, up to ~ 15 kOe [Vic05]; thus the nano-wire system presented here can cope with these values. However, next steps in further developing the system suggested here should also concentrate on simulations and experiments with the specific alloys used in the large hard disk producing companies.

With these technical parameters, the quaternary magnetic nano-particles investigated in this thesis can be the base for novel developments in magnetic storage systems, allowing for furthermore keeping pace with Moore's law in spite of the physical and technical limits which at present slow down the progress in hard disk drive enhancements.

Literature

[Ake05] J. Akerman: Towards a Universal Memory, Science **308**, 508-510 (2005)

[Ama10] E. Amaladass, B. Ludescher, G. Schütz, T. Tyliszczak, M.-S. Lee, and T. Eimüller: Nanospheres generate out-of-plane magnetization, J. Appl. Phys. **107**, 053911 (2010)

[Ame10] A. Amer, D. D. E. Long, E. L. Miller, J.-F. Paris, T. Schwarz: Design Issues for a Shingled Write Disk System, Proc. of 26th IEEE (MSST2010) Symposium on Massive Storage Systems and Technologies, Incline Village, NV / USA, May 3-7, 2010

[Arg08] A. J. Argumendo, D. Berman, R. G. Biskeborn, G. Cherubini, R. D. Cideciyan, E. Eleftheriou, W. Häberle, D. J. Hellman, R. Hutchins, W. Imaino, J. Jelitto, K. Judd, P.-O. Jubert, M. A. Lantz, G. M. McClelland, T. Mittelholzer, C. Narayan, S. Ölçer, P. J. Seger: Scaling Tape-Recording Areal Densities to 100 Gb/in², IBM J. Res. Develop. **52**, 513-527 (2008)

[Asa97] A. Asamitsu, Y. Tomioka, H. Kuwahara, Y. Tokura: Current-Switching of Resistive States in Colossal Magnetoresistive Oxides, Nature **388**, 50-52 (1997)

[Bad06] S. D. Bader: Opportunities in nanomagnetism, Rev. Mod. Phys. **78**, 1 (2006)

[Bec98] S. Becker: Feingebrannt – Höhere Speicherdichten bei magnetooptischen Wechselplatten, c't 25/1998, 190 (1998)

[Bec00] A. Beck, J. G. Bednorz, C. Gerber, C. Rossel, D. Widmer: Reproducible Switching Effect in Thin Oxide Films for Memory Applications, Appl. Phys. Lett. **77**, 139-141 (2000)

[Bec03] B. Beckmann, U. Nowak, and K. D. Usadel: Asymmetric Reversal Modes in Ferromagnetic/Antiferromagnetic Multilayers, Phys. Rev. Lett. **91**, 187201 (2003)

[Bed07] D. Bedau, M. Kläui, S. Krzyk, U. Rüdiger, G. Faini, L. Vila: Detection of Current-Induced Resonance of Geometrically Confined DomainWalls, Phys. Rev. Lett. **99**, 146601 (2007)

[Ber96] L. Berger: Emission of spin waves by a magnetic multilayer traversed by a current, Phys. Rev. B **54**, 9353-9358 (1996)

[Bla07] T. Błachowicz, A. Tillmanns, M. Fraune, B. Beschoten, and G. Güntherodt: Exchange-bias in (110)-oriented CoO/Co bilayers with different magnetocrystalline anisotropies, Phys. Rev. B **75**, 054425 (2007)

[Bla07a] T. Blachowicz: Rotatable anisotropy in epitaxial exchange-biased materials revealed by Brillouin light scattering, J. Appl. Phys. **102**, 043901 (2007)

[Bla10] T. Blachowicz, A. Ehrmann neé Tillmanns, P. Steblinski, L. Pawela: Magnetization reversal in magnetic half-balls influenced by shape perturbations, J. Appl. Phys. **108**, 123906 (2010)

[Bla11] T. Blachowicz, A. Ehrmann: Fourfold nanosystems for quaternary storage devices, J. Appl. Phys. **110**, 073911 (2011)

[Bla12] T. Blachowicz, A. Ehrmann: Anatomy of Demagnetizing and Exchange Fields in Magnetic Nanodots Influenced by 3D Shape Modifications, arXiv:1207.4673v1 (2012)

[Bla13a] T. Blachowicz, A. Ehrmann: Six-state, three-level, six-fold ferromagnetic wire system, J. Magn. Magn. Mat. **331**, 21-23 (2013)

[Bla13b] T. Blachowicz, A. Ehrmann, P. Steblinski, J. Palka: Directional-dependent coercivities and magnetization reversal mechanisms in fourfold ferromagnetic systems of varying sizes, J. Appl. Phys. **113**, 013901 (2013)

[Bor10] J. Borghetti, G. S. Snider, P. J. Kuekes, J. J. Yang, D. R. Stewart, and R. S. Williams: "Memristive" Switches enable 'Stateful' Logic Operations via Material Implication, Nature **464**, 873-876 (2010)

[Bow09] S. R. Bowden and U. J. Gibson: Optical Characterization of All-Magnetic NOT Gate Operation in Vortex Rings, IEEE Trans. Magn. **45**, 5326 (2009)

[Bro11] G. Brown: How Floppy Disk Drives Work, HowStuffWorks (http://www.howstuffworks.com/floppy-disk-drive2.htm)

[Cag11] C. Cagli, F. Nardi, B. Harteneck, Z. Tan, Y. Zhang, and D. Ielmini: Resistive-switching crossbar memory based on Ni-NiO core-shell nanowires, Small **7**, 2899-2905 (2011)

[Car08] M. J. Carey, S. Maat, N. Smith, R. E. E. Fontana, Jr., D. Druist, K. J. Carey, J. A. Katine, N. Robertson, T. D. Boone, Jr. M. Alex, J. O. Moore, C. H. Tsang: All-Metal Current-Perpendicular-to-Plane Giant Magnetoresistance Sensors for Narrow-Track Magnetic Recording, IEEE Trans. Magn. **44**, 90-94 (2008)

[Che06] Y. Chen, F. X. Liu, X. Chen, B. Xu, P.-L. Lu, M. S. Patwari, H. Xi, C. H. Chang, B. Miller, D. Ménard, B. B. Pant, J. Loven, K. Duxstad, S. Li, Z. Zhang, S. B. Jonston, R. W. Lamberton, M. A. Gubbins, T. K. McLaughlin, J. B. Gadbois, J. Ding, B. Cross, S. S. Xue, P. J. Ryan: Commercial TMR heads for hard disk drives: characterization and extendibility at 300 gbit/in², IEEE Trans. Magn. **42**, 97-102

[Che10] G. Cherubini, R. D. Cideciyan, L. Dellmann, E. Eleftheriou, W. Häberle, J. Jelitto, V. Kartik, M. A. Lantz, S. Ölcer, A. Pantazi, H. E. Rothuizen, D. Berman, W. Imaino, P.-O. Jubert, G. McClelland, P. V. Köppe, K. Tsuruta, T. Harasawa, Y. Murata, A. Musha, H. Noguchi, H. Ohtsu, S. Shimizu, R. Suziki: 29.5 Gb/in² Recording Areal Density on Barium Ferrite Tape, IEEE Trans. Magn. **47**, 137-144 (2011)

[Cho07] K. W. Chou, A. Puzic, H. Stoll, D. Dolgos, G. Schütz, B. Van Waeyenberge, A. Vansteenkiste, T. Tyliszczak, G. Woltersdorf, and C. H. Back: Direct observation of the vortex core magnetization and its dynamics, Appl. Phys. Lett. **90**, 202505 (2007)

[Chu10] S.-W. Chung, K.-M. Rho, S. D. Kim, H.-J. Suh, D.-J. Kim, H.-J. Kim, S.-H. Lee, J.-H. Park, H.-M. Hwang, S.-M. Hwang, J.-Y. Lee, Y.-B. An, J.-U. Yi, Y.-H. Seo, D.-H. Jung, M.-S. Lee, S.-H. Cho, J.-N. Kim, G.-J. Park, J. Gyuan; A. Driskill-Smith, V. Nikitin, A. Ong, X. Tang, K. Yongki, J.-S. Rho, S.-K. Park, S.-W. Chung, J.-G. Jeong, S.-J. Hong: Fully integrated 54 nm STT-RAM with the smallest bit cell dimension for high density memory application, 2010 IEEE Int. Electron Devices Meeting (San Francisco, CA), pp. 12.7.1-12.7.4 (2010)

[Cla04] J. Clarke, A. I. Braginski (Ed.): The SQUID Handbook Vol. I: Fundamentals and Technology of SQUIDs and SQUID Systems, Wiley-VCH 2004

[Cou12] T. Coughlin, E. Grochowski: Years of destiny: HDD capital spending and technology developments from 2012-2016, IEEE Magnetics Society Meeting, June 19, 2012

[Cow99] R. P. Cowburn, D. K. Koltsov, A. O. Adeyeye, M. E. Welland, and D. M. Tricker: Single-Domain Circular Nanomagnets, Phys. Rev. Lett. **83**, 1042 (1999)

[Cow00] R. P. Cowburn and M. E. Welland: Room temperature magnetic quantum cellular automata, Science **287**, 1466 (2000)

[Cow02] R. P. Cowburn, D. A. Allwood, G. Xiong, and M. D. Cooke: Domain wall injection and propagation in planar Permalloy nanowires, J. Appl. Phys. **91**, 6949 (2002)

[Cro06] G. L. W. Cross: The Production of Nanostructures by Mechanical Forming, J. Phys. D: Appl. Phys. **39**, R363-386 (2006).

[Dan10] A. L. Dantas, G. O. G. Rebouças, and A. S. Carriço: Vortex Nucleation in Exchange Biased Magnetic Nanoelements, IEEE Trans. Magn. **46**, 2311 (2010)

[Dia10] Z. Diao, M. Abid, P. Upadhyaya, M. Venkatesan, and J. M. D. Coey: Vortex states in soft magnets in two and three dimensions, J. Magn. Magn. Mat. **322**, 1304 (2010)

[Don08] Y. Dong, G. Yu, M. C. McAlpine, W. Lu, and C. M. Lieber: Si/a-Si Core/Shell Nanowires as Nonvolatile Crossbar Switches, Nano Lett. **8**, 386-391 (2008)

[Ehr11] A. Ehrmann, T. Błachowicz: Adjusting exchange bias and coercivity of magnetic layered systems with varying anisotropies, J. Appl. Phys. **109**, 083923 (2011)

[Ehr11a] A. Ehrmann, T. Błachowicz, P. Steblinski, M. O. Weber: Neue Anisotropien - von der Grundlagenforschung zu optimierten magnetischen Speichermedien, in A. Brenke (Ed.): ASIM-Konferenz STS/GMMS 2011, Proceedings, Shaker Verlag (2011)

[Ele10] E. Eleftheriou, R. Haas, J. Jelitto, M. Lantz, and H. Pozidis: Trends in Storage Technologies, IEEE Data Eng. Bull. **33**, 4-13 (2010)

[Elt10] M. Eltschka, M. Wötzel, J. Rhensius, S. Krzyk, U. Nowak, M. Kläui, T. Kasama, R. E. Dunin-Borkowski, L. J. Heyderman, H. J. van Driel, R. A. Duine: Nonadiabatic Spin Torque Investigated Using Thermally Activated Magnetic DomainWall Dynamics, Phys. Rev. Lett. **105**, 056601 (2010)

[Esc13] Hard Drives, Escotal.com Computer Training – Internet Consulting (http://www.escotal.com/harddrive.html)

[Fas96] J. Fassbender: Struktur und Magnetismus von Co-Schichten auf Cu-Einkristallen, dissertation thesis, RWTH Aachen 1996

[Fer99] A. Fert and L. Piraux : Magnetic nanowires, J. Magn. Magn. Mater. **200**, 338 (1999)

[Fon11] M. Fonin, C. Hartung, U. Rüdiger, D. Backes, L. Heyderman, F. Nolting, A. Fraile Rodríguez, and M. Kläui: Formation of magnetic domains and domain walls in epitaxial Fe3O4(100) elements (invited), J. Appl. Phys. **109**, 07D315 (2011)

[Fon13] R. E. Fontana, Jr., G. M. Decad, S. R. Hetzler: The Impact of Areal Density and Millions of Square Inches (MSI) of Produced Memory on Petabyte Shipments of TAPE, NAND Flash, and HDD Storage Class Memories, 29th IEEE Conference on Massive Data Storage, Long Beach, May 6-10, 2013

[Fuc05] G. D. Fuchs, I. N. Krivorotov, P. M. Braganca, N. C. Emley, A. G. F. Garcia, D. C. Ralph, R. A. Buhrmann: Adjustable spin torque in magnetic tunnel junctions with two fixed layers, Appl. Phys. Lett. **86**, 152509 (2005)

[Fuj06] Fujitsu Press Release: Fujitsu and Tokyo Institute of Technology Announce the Development of New Material for 256Mbit FeRAM Using 65-nanometer Technology – Low Power and High Speed FeRAMs for New Mobile Electronic Products, August 8, 2006 (http://www.fujitsu.com/sg/news/pr/fmal_20060808.html)

[Gaj12] M. Gajek, J. J. Nowak, Z. Z. Sun, P. L. Trouilloud, E. J. O'Sullivan, D. W. Abraham, M. C. Gaidis, G. Hu, S. Brown, Y. Zhu, R. P. Robertazzi, W. J. Gallagher, and D. C. Worledge: Spin torque switching of 20 nm magnetic tunnel junctions with perpendicular anisotropy, Appl. Phys. Lett. **100**, 132408 (2012)

[Gao07] X. S. Gao, A. O. Adeyeye, S. Goolaup, N. Singh, W. Jung, F. J. Castaño, and C. A. Ross: Inhomogeneities in spin states and magnetization reversal of geometrically identical elongated Co rings, J. Appl. Phys. **101**, 09F505 (2007)

[GiD13] GID – The Personal Pre And Post Processor (http://gid.cimne.upc.es/)

[Gon13] J. B. González-Díaz, J. A. Arregi, A. Martínez-de-Guerenu, F. Arizti, and A. Berger: Quantitative magneto-optical characterization of diffusive reflected light from rough steel samples, J. Appl. Phys. **113**, 153904 (2013)

[Gre09] S. Greaves, Y. Kanai, and H. Muraoka: Shingled recording for 2-3 Tbit/in², IEEE Trans. Magn. **45**, 3823-2829 (2009)

[Gri04] M. Grimsditch and P. Vavassori: The diffracted magneto-optic Kerr effect: what does it tell you?, J. Phys.: Concens. Matter **16**, R275 (2004)

[Gro10] M. Grobis, E. Dobisz, O. Hellwig, M. E. Schabes, G. Zeltzer, T. Hauet, and T. R. Albrecht: Measurements of the write error rate in bit patterned magnetic recording at 100-320 Gb/in², Appl. Phys. Lett. **96**, 052409 (2010)

[Gus02] K. Yu. Guslienko, B. A. Ivanov, V. Novosad, Y. Otani, H. Shima, and K. Fukamichi: Eigenfrequencies of vortex state excitations in magnetic submicron-size disks, J. Appl. Phys. **91**, 8037 (2002)

[Gus06] K. Y. Guslienko, X. F. Han, D. J. Keavney, R. Divan, and S. D. Bader: Magnetic vortex core dynamics in cylindrical ferromagnetic dots, Phys. Rev. Lett. **96**, 067205 (2006)

[Hay08] M. Hayashi, L. Thomas, R. Moriya, C. Rettner, S. S. P. Parkin: Current-Controlled Magnetic Domain-Wall Nanowire Shift Register, Science **320**, 209-211 (2008)

[He10] K. He, D. J. Smith, and M. R. McCartney: Effects of vortex chirality and shape anisotropy on magnetization reversal of Co nanorings (invited), J. Appl. Phys. **107**, 09D307 (2010)

[Hen01] Y. Henry, K. Ounadjela, L. Piraux, S. Dubois, J.-M. George, J.-L. Duvail: Magnetic anisotropy and domain patterns in electrodeposited cobalt nanowires, Eur. Phys. J. B **20**, 35 (2001)

[Her01] R. Hertel: Micromagnetic simulations of magnetostatically coupled Nickel nanowires, J. Appl. Phys. **90**, 5752 (2001)

[Her07] R. Hertel, S. Gliga, M. Fähnle, and C. M. Schneider: Ultrafast Nanomagnetic Toggle Switching of Vortex Cores, Phys. Rev. Lett. **98**, 117201 (2007)

[Her09] E. P. Hernández, A. Azevedo, and S. M. Rezende: Structure and magnetic properties of hexagonal arrays of ferromagnetic nanowires, J. Appl. Phys. **105**, 07B525 (2009)

[Hie97] W. K. Hiebert, A. Stankiewicz, and M. R. Freeman: Direct observation of magnetic relaxation in a small permalloy disk by time-resolved scanning Kerr microscopy, Phys. Rev. Lett. **79**, 1134 (1997)

[Hie03] W. K. Hiebert, L. Lagae, and J. De Boeck: Spatially inhomogeneous ultrafast precessional magnetizition reversal, Phys. Rev. B **68**, 020402 (2003)

[Hil90] B. Hillebrands: Spin-wave calculations for multilayered structures, Phys. Rev. B **41**, 530 (1990)

[Hit10] Hitachi Global Storage Technologies – Hitachi Research and Technology – Overview (http://www1.hgst.com/hdd/research/)

[Hua01] Y. Huang, X. Duan, Q. Wei, C. M. Lieber: Directed Assembly of One-Dimensional Nanostructures into Functional Networks, Science **291**, 630 (2001)

[Hua01a] Y. Huang, X. Duan, Y. Cui, L. J. Lauhon, K.-H. Kim, C. M. Lieber: Logic Gates and Computation from Assembled Nanowire Building Blocks, Science **294**, 1313 (2001)

[Hua10] L. Huang, M. A. Schofield, and Y. Zhu: Control of Double-Vortex Domain Configurations in a Shape-Engineered Trilayer Nanomagnet System, Adv. Mater. **22**, 492 (2010)

[IBM55] International Business Machines Corporation: The 650 Magnetic Drum Data Processing Machine (http://archive.computerhistory.org/resources/text/ IBM/IBM.650.1955.102646125.pdf)

[IBM56] IBM 350 disk storage unit, IBM Archives http://www-03.ibm.com/ibm/history/exhibits/storage/storage_350.html)

[IBM71] IBM 100 – The Floppy Disk (http://www-03.ibm.com/ibm/history/ibm100/us/en/icons/floppy/breakthroughs/)

[IBM96] Did you ever wonder how your hard disk drive works?, IBM Research (http://www.research.ibm.com/research/gmr/basics.html)

[IBM12] Atomic-Scale Magnetic Memory, IBM / Smarter Computing (http://www.ibm.com/smarterplanet/global/files/us__en_us__computing__atomic_scale_magnetic_memory_011212.pdf)

[IBM13] GMR – The Giant Magnetoresistive Head: A giant leap for IBM Research (http://www.research.ibm.com/research/gmr.html)

[IBM13a] Magnetic Stripe Technology, IBM100 – Icons of Progress (http://www-03.ibm.com/ibm/history/ibm100/us/en/icons/magnetic/)

[Iel11] D. Ielmini, F. Nardi and C. Cagli: Physical models of size-dependent nanofilament formation and rupture in NiO resistive switching memories, Nanotechnology **22**, 254022 (2011)

[Iel13] D. Ielmini, C. Cagli, F. Nardi, and Y. Zhang: Nanowire-based resistive switching memories: deviced, operation and scaling, J. Phys. D: Appl. Phys. **46**, 074006 (2013)

[ITRS] ITRS Roadmap (www.itrs.net)

[Jeo98] W.-Ch. Jeong, B.-I. Lee, and S.-K. Joo: A new multilayered structure for multilevel magnetoresistive random access memory (MRAM) cell, IEEE Trans. Magn. **34**, 1069-1071 (1998)

[Jeo99] W.-Ch. Jeong, B.-I. Lee, and S.-K. Joo: Three level, six state multilevel magnetoresistive RAM(MRAM), J. Appl. Phys. **85**, 4782 (1999)

[Jon76] J. R. Jones: Coincident Current Ferrite Core Memories, Byte magazine, July 1976

[Kar11] G. Kartopu, O. Yalçın, K.-L. Choy, R. Topkaya, S. Kazan, and B. Aktaş: Size effects and origin of easy-axis in nickel nanowire arrays, J. Appl. Phys. **109**, 033909 (2011)

[Kas06] S. Kasai, Y. Nakatani, K. Kobayashi, H. Kohno, and T. Ono : Current-driven resonant excitation of magnetic vortices, Phys. Rev. Lett. **97**, 107(2006)

[Kaw12] T. Kawahara, K. Ito, R. Takemura, H. Ohno: Spin-transfer torque RAM technology : Review and prospect, Microelectronics Reliability **52**, 613-627 (2012)

[Khv12] A. V. Khvalkovskiy, D. Apalkov, S. Watts, R. Chepulskii, R. S. Beach, A. Ong, X. Tang, A. Driskill-Smith, W. H. Butler, P. B. Visscher, D. Lottis, E. Chen, V. Nikitin, and M. Krounbi: Basic principles of STT-MRAM cell operation in memory arrays, J. Phys. D: Appl. Phys. **46**, 074001 (2013)

[Kim10a] J.-S. Kim, O. Boulle, S. Verstoep, L. Heyne, J. Rhensius, M. Kläui, L. J. Heyderman, F. Kronast, R. Mattheis, C. Ulysse, G. Faini: Current-induced vortex dynamics and pinning potentials probed by homodyne detection, Phys. Rev. B **82**, 104427 (2010)

[Kim10] R. Kim, S. Holmes, S. Halle, V. Dai, J. Meiring, A. Dave, M. E. Colburn, and H. J. Levinson: 22nm technology node active layer patterning for planar transistor devices, J. Micro/Nanolith. MEMS MOEMS **9**, 013001 (2010)

[Klä03] M. Kläui, C. A. F. Vaz, J. A. C. Bland, T. L. Monchesky, J. Unguris, E. Bauer, S. Cherifi, S. Heun, A. Locatelli, L. J. Heyderman, Z. Cui: Direct observation of spin configurations and classification of switching processes in mesoscopic ferromagnetic rings, Phys. Rev. B **68**, 134426 (2003)

[Kne91] E. F. Kneller and R. Hawig: The exchange-spring magnet: A new material principle for permanent magnets, IEEE Trans. Magn. **27**, 3588 (1991)

[Kod99] R. H. Kodama: Magnetic nanoparticles, J. Magn. Magn. Mat. **200**, 359 (1999)

[Kon12] J.-I. Kon, T. Maruyama, Y. Kojima, Y. Takahashi, S. Sugatani, K. Ogino, H. Hoshino, H. Isobe, M. Kurokawa, A. Yamada: Optimization of chemically amplified resist for high-volume manufacturing by electron-beam direct writing toward 14nm node and beyond, Proc. SPIE **8323**, Alternative Lithographic Technologies IV, 832324 (2012)

[Koz99] M. N. Kozicki, M. Yun, L. Hilt, A. Singh: Applications of Programmable Resistance Changes in Metal-Doped Chalconides, Electrochem. Soc. Proc. **99-13**, 298-309 (1999)

[Lem03] Telegraphone, Inventor of the Week, Lemelson-MIT Program (http://web.mit.edu/invent/iow/poulsen.html)

[Leo10] T. G. Leong, A. M. Zarafshar, and D. H. Gracias: Three-Dimensional Fabrication at Small Size Scales, Small **6**, 792 (2010)

[Li09] H. Li, Y. Chen: An overview of non-volatile memory technology and the implication for tools and architectures, Design, Automation & Test in Europe Conference & Exhibition '09 (2009)

[Li10] Y. Li, T. X. Wang, and Y. X. Li: The influence of dipolar interaction on magnetic properties in nanomagnets with different shapes, Phys. Stat. Sol. b **247**, 1237 (2010)

[Li12] D. Li, J. S. Vetter, G. Marin, C. McCurdy, C. Cira, Z. Liu and W. Yu: Identifying opportunities for byte-addressable non-volatile memory in extreme-scale scientific applications, in: IPDPS, pp. 945-956, IEEE Computer Society (2012)

[Li13] T. Li, A. Patz, L. Mouchliadis, J. Yan, T. A. Lograsso, I. E. Perakis, and J. Wang: Femtosecond switching of magnetism via strongly correlated spin-charge quantum excitations, Nature **496**, 69-73 (2013)

[Liu12] L. Liu, C.-F. Pai, Y. Li, H. W. Tseng, D. C. Ralph, R. A. Buhrman: Spin-torque switching with the giant spin hall effect of tantalum, Science **336**, 555-558 (2012)

[mag13] magpar – Parallel Finite Element Micromagnetics Package (http://www. magpar.net/ static/magpar/doc/html/index.html)

[Mak11] A. Makarov, V. Sverdlov, D. Osintsev, S. Selberherr: Reduction of switching time in pentalayer magnetic tunnel junctions with a composite-free layer, Phys. Status Solidi RRL **5**, 1-3 (2011)

[Mak11a] A. Makarov, V. Sverdlov, D. ÄOsintsev, S. Selberherr: Fast switching in magnetic tunnel junctions with double barrier layer, Proc. of 2011 SSDM, 456-457 (2011)

[Mak12] A. Makarov, V. Sverdlov, S. Selberherr: Emerging memory technologies: Trends, challenges, and modeling methods, Microelectronics Reliability **52**, 628-634 (2012)

[Mar09] T. Maruyama, Y. Shiota, T. Nozaki, K. Ohta, N. Toda, M. Mizuguchi, A. A. Tulapurkar, T. Shinjo, M. Shiraishi, S. Mizukami, Y. Ando, and Y. Suzuki: Large voltage-induced magnetic anisotropy change in a few atomic layers of iron, Nature Nanotechnol. **4**, 158-161 (2009)

[Mar12] T. Maruyama, Y. Machida, S. Sugatani, H. Takita, H. Hoshino, T. Hino, M. Ito, A. Yamada, T. Iizuka, S. Komatsu, M. Iked, K. Asada: CP element based design for 14 nm node EBDW high volume manufacturing, Proc. SPIE **8323**, Alternative Lithographic Technologies IV, 832314 (2012)

[McM98] R.D. McMichael, M.D. Stiles, P.J. Chen, and W.F. Egelhoff Jr: Ferromagnetic resonance studies of NiO-coupled thin films of Ni80Fe20, Phys. Rev. B **58**, 8605 (1998)

[Mei56] W. H. Meiklejohn and C. P. Bean (1956): New magnetic anisotropy, Phys. Rev. **102**, 1413

[Mei65] R. H. Meinken and L. W. Stammerjohn: Memory Devices, Bell Laboratories Record **49**(6), June 1965

[Mei07] G. Meier, M. Bolte, R. Eiselt, B. Krüger, D.-H. Kim, P. Fischer: Direct Imaging of Stochastic Domain-Wall Motion Driven by Nanosecond Current Pulses, Phys. Rev. Lett. **98**, 187202 (2007)

[Mej10] J. Mejía-López, D. Altbir, P. Landeros, J. Escrig, A. H. Romero, Igor V. Roshchin, C.-P. Li, M. R. Fitzsimmons, X. Batlle, and I. K. Schuller: Development of vortex state in circular magnetic nanodots: Theory and experiment, Phys. Rev. B **81**, 184417 (2010)

[MIC13] MICR – Magnetic Ink Character Recognition (http://www.whatismicr.com/index.html)

[Moo65] G. E. Moore: Cramming more components onto integrated circuits, Electronics Magazine April 1965, p. 4 (1965)

[Mor11] J. Moritz, G. Vinai, S. Auffret, and B. Dieny: Two-bit-per-dot patterned media combining in-plane and perpendicular-to-plane magnetized thin films, J. Appl. Phys. **109**, 083902 (2011)

[Mül03] G. Müller, N. Nagel, C.-U. U. Pinnow, T. Röhr: Emerging Non-Volatile Memory Technologies, Proc. of the 29th European Solid-State Circuits Conference, ESSCIRC '03 (2003)

[Mun06] M. Muneeb, I. Akram, A. Nazir: Smart Electronic Materials – Non-Volatile Random Access Memory Technologies, Lecture, KTH – School of information and communication technology, Sweden 2006 (http://citeseerx.ist.psu.edu/viewdoc/summary?doi=10.1.1.98.2239)

[Nav78] Thin film, Digital Computer Basics, Naval Education and Training Command, rev. 1978; cited from: E. Thelen: Facts and stories about Antique (lonesome) Computers (http://ed-thelen.org/comp-hist/navy-thin-film-memory-desc.html)

[NCR62] NCR CRAM: Card Random Access Memory, product brochure, The National Cash Register Co. (http://archive.computerhistory.org/resources/text/ NCR/NCR.CRAM.1960.102646240.pdf)

[Nie01] K. Nielsch, R. B. Wehrspohn, J. Barthel, J. Kirschner, U. Gösele, S.F. Fischer, and H. Kronmüller: Hexagonally Ordered 100 nm Period Nickel Nanowire Arrays, Appl. Phys. Lett. **79**, 1360 (2001)

[Nog05] J. Nogues, J. Sort, V. Langlais, V. Skumryev, S. Surinach, J. S. Munoz, and M. D. Baro: Exchange bias in nanostructures, Phys. Rep. 422, 65 (2005)

[Nov05] V. Novosad, F. Fradin, P. Roy, K. Buchanan, K. Guslienko, S. D. Bader: Magnetic vortex resonance in patterned ferromagnetic dots, Phys. Rev. B 72, 024455 (2005)

[Ogo12] K. Ogino, H. Hoshino, T. Maruyama, Y. Machida, and S. Sugatani: Proximity effect correction using multilevel area density maps for character projection based electron beam direct writing toward 14 nm node and beyond, Proc. SPIE **8323**, Alternative Lithographic Technologies IV, 832328 (2012)

[Ohn11] H. Ohno: Magnetoresistive random access memory with spin transfer torque write (Spin RAM) – present and future, Proc. of 2011 SSDM, 957-958 (2011)

[Ols59] K. H. Olsen, R. L. Best: Magnetic core memory, US Patent 3161861 (1959)

[Ou11] E. Ou, P. Leong: Emerging non-volatile memory technologies for reconfigurable architectures, 2011 IEEE 54th International Midwest Symposium on Circuits and Systems (MWSCAS)

[PCG05] Function of the Read/Write Heads, The PC Guide, last updated 2005 (http://www.pcguide.com/ref/hdd/op/heads/opFunction-c.html)

[Par08] S. S. P. Parkin, M. Hayashi, L. Thomas: Magnetic Domain-Wall Racetrack Memory, Science **320**, 190-194 (2008)

[Par13] ParaView – Open Source Scientific Visualization (http://paraview.org/)

[Pir07] S. N. Piramanayagam: Perpendicular recording media for hard disk drives, J. Appl. Phys. **102**, 011301 (2007)

[Pit11] K. Pitzschel, J. Bachmann, S. Martens, J. M. Motero-Moreno, J. Kimling, G. Meier, J. Escrig, K. Nielsch, and D. Görlitz: Magnetic reversal of cylindrical nickel nanowires with modulated diameters, J. Appl. Phys. **109**, 033907 (2011)

[PTC13] PTC – PTC Mathcad – Engineering Calculations Software (http://www.ptc.com/ product/mathcad/)

[Pul10] J. F. Pulecio, S. Bhanja: Magnetic cellular automata coplanar cross wire systems, J. Appl. Phys. **107**, 034308 (2010)

[Qua13] Quantum LTO-6/LTO-5 Tape Drive Datasheet (https://iq.quantum.com/exLink.asp?10444458OP44N16I37407297)

[Qua13a] Quantum LTO-5 Tape Drive User's Guide (http://downloads.quantum.com/lto5/6-66786-01_RevA.pdf)

[Red10] C. Redondo, B. Sierra, S. Moralejo, and F. Castano: Magnetization reversal induced by irregular shape nanodots in square arrays, J. Magn. Magn. Mat. **322**, 1969 (2010)

[Rem08] A. Remhof, A. Schumann, A. Westphalen, H. Zabel, N. Mikuszeit, E. Y. Vedmedenko, T. Last, and U. Kunze: Magnetostatic interactions on a square lattice, Phys. Rev. B **77**, 134409 (2008)

[Ric06] H. Richter, A. Dobin, O. Heinonen, K. Gao, R. Veerdonk, R. Lynch, J. Xue, D. Weller, P. Asselin, M. Erden, R. Brockie: Recording on bit-patterned media at densities of 1 Tb/in² and beyond, IEEE Trans. Magn. **42**, 2255-2260 (2006)

[Ros01] C. A. Ross: Patterned magnetic recording media, Annu. Rev. Mater. Res. **31**, 203 (2001)

[Ros07] G. Rostky: Bubbles: the better memory, 2000 – The Century of the Engineer – Misunderstood Milestones, EETimes.com (http://web.archive.org/web/ 20070930031657/http://www.eetonline.com/special/special_issues/mi llennium/milestones/bobeck.html)

[Rot01] J. Rothman, M. Kläui, L. Lopez-Diaz, C. A. F. Vaz, A. Bleloch, J. A. C. Bland, Z. Cui, R. Speaks: Observation of a Bi-Domain State and Nucleation Free Switching in Mesoscopic Ring Magnets, Phys. Rev. Lett. **86**, 1098 (2001)

[Sbi11] R. Sbiaa, S. Y. H. Lua, R. Law, H. Meng, R. Lye, and H. K. Tan: Reduction of switching current by spin transfer torque effect in perpendicular anisotropy magnetoresistive devices (invited), J. Appl. Phys. **109**, 07C707 (2011)

[Sbi11a] R. Sbiaa, H. Meng, S. N. Piramanayagam: Materials with perpendicular magnetic anisotropy for magnetic random access memory, Phys. Status Solidi RRL **5**, 413-419 (2011)

[Sch03] W. Scholz, J. Fidler, T. Schrefl, D. Suess, R. Dittrich, H. Forster, V. Tsiantos: Scalable Parallel Micromagnetic Solvers for Magnetic Nanostructures, Comp. Mat. Sci. **28**, 366 (2003)

[Sch03a] W. Scholz: Scalable Parallel Micromagnetic Solvers for Magnetic Nanostructures, PhD Thesis, p. 161 (2003)

[Sea13] Seagate: Desktop Hard Drive SATA 6Gb/s 4TB HDD Capacity (http://www.seagate.com/internal-hard-drives/desktop-hard-drives/desktop-hdd/#)

[Sel77] J. E. Seleznev, J. A. Burkin, S. V. Kuzmin: Ferrite core memory, US Patent 4161037 (1977)

[Sil06] E. L. Silva, W. C. Nunes, M. Knobel, J.C. Denardin, D. Zanchet, K. Pirota, D. Navas, M. Vázquez: Transverse magnetic anisotropy of magnetoelastic origin induced in Co nanowires, Physica B **384**, 22 (2006)

[Slo96] J. C. Slonczewski: Current-driven excitation of magnetic multilayers, J. Magn. Magn. Mater. **159**, L1-7 (1996)

[Slo02] J. C. Slonczewski: Currents and torques in metallic magnetic multilayers, J. Mag. Magn. Mat. **247**, 324-338 (2002)

[Smi89] N. Smith, D. Markham, and D. LaTourette: Magnetoresistive measurement of the exchange constant in varied-thickness permalloy films, J. Appl. Phys. **65**, 4362 (1989)

[Smi05] T. Smith: Seagate pledges first 2.5in perpendicular HDD, The Register 8th June 2005 (http://www.theregister.co.uk/2005/06/08/seagate_hdd_roadmap/)

[Soa08] M. M. Soares, E. de Biasi, L. N. Coelho, M. C. dos Santos, F. S. de Menezes, M. Knobel, L. C. Sampaio, and F. Garcia: Magnetic vortices in tridimensional nanomagnetic caps observed using transmission electron microscopy and magnetic force microscopy, Phys. Rev. B **77**, 224405 (2008)

[Sti99] M.D. Stiles and R.D. McMichael: Model for Exchange Bias in Polycrystalline Ferromagnet-Antiferromagnet Bilayers, Phys. Rev. B **59**, 3722 (1999)

[Sub04] A. Subramani, D. Geerpuram, A. Domanowski, V. Baskaran, and V. Metlushko: Vortex state in magnetic rings, Physica C **404**, 241 (2004)

[Tag01] I. Tagawa, S. Ikeda, Y. Uehara: High-Performance Write Head Design and Materials, FUJITSU Sci. Tech. J. 37, 164-173 (2001)

[Tan19] D. D. Tang, Y.-J. Lee: Magnetic Memory: fundamentals and technology, Cambridge University Press 2010

[TDK13] Magnetic head technology paved the way for smaller hard disk drives with higher storage capacities, The Wonders of Electromagnetism Vol. 2, TDK Tech-Mag 2013.01.31 (http://www.global.tdk.com/techmag/inductive/vol2/index3.htm)

[Teh00] S. Tehrani, B. Engel, J. Slaughter, E. Y. Chen, M. DeHerrera, M. P. Naji, R. Whig, J. Janesky, and J. Calder: Recent developments in Magnetic Tunnel Junction MRAM, IEEE Trans. Magn. **36**, 2752 (2000)

[Ter05] B. D. Terris and T. Thomson: Nanofabricated and self-assembled magnetic structures as data storage media, J. Phys. D: Appl. Phys. **38**, R199 (2005)

[The10] L. Thevenard, H. T. Zeng, D. Petit, R. P. Cowburn: Macrospin limit and configurational anisotropy in nanoscale permalloy triangles, J. Magn. Magn. Mater. **322**, 2152 (2010)

[Thi12] N. Thiyagarajah, K.-I. Lee, and S. Bae: Spin transfer switching characteristics in a $[Pd/Co]_m/Cu/[Co/Pd]_n$ pseudo spin-valve nanopillar with perpendicular anisotropy, J. Appl. Phys. **111**, 07C910 (2012)

[Til01] A. Tillmanns: Abhängigkeit von Exchange-Bias- und Koerzitivfeld von der Orientierung, Verdünnung und Stapelfolge im Schichtsystem Co/CoO, Diploma Thesis, RWTH Aachen 2001

[Til06] A. Tillmanns, S. Oertker, B. Beschoten, G. Güntherodt, C. Leighton, I. K. Schuller, J. Nogués: Magneto-optical study of magnetization reversal asymmetry in exchange bias, Appl. Phys. Lett. **89**, 202512 (2006)

[Til06a] A. Tillmanns: Magnetisierungsumkehr und -dynamik in Exchange-Bias-Systemen, Dissertation Thesis, RWTH Aachen 2006

[Til09] A. Tillmanns, T. Błachowicz, M. Fraune, G. Güntherodt, and I. K. Schuller: Anomalous magnetization reversal mechanism in unbiased Fe/FeF2 investigated by means of the magneto-optic Kerr effect, J. Magn. Magn. Mater. **321**, 2932-2935 (2009)

[Tos04] Toshiba Leads Industry in Bringing Perpendicular Data Recording to HDD – Sets New Record for Storage Capacity With Two New HDDs, 14 Dec., 2004
(http://www.toshiba.co.jp/about/press/2004_12/pr1401.htm)

[Tos10] Bit-Patterned Media for High-Density HDDs, TOSHIBA Technologies, LSI & Storage
(http://www.toshiba.co.jp/rdc/rd/fields/11_e09_e.htm)

[Tud10] I. Tudosa, J. A. Katine, S. Mangin, and Eric E. Fullerton: Perpendicular spin-torque switching with a synthetic antiferromagnetic reference layer, Appl. Phys. Lett. **96**, 212504 (2010)

[Ult13] Ultrium – LTO Technology – Ultrium GenerationsLTO
(http://www.lto-technology.com/technology/generations.html)

[Uso07] N. A. Usov, A. Zhukov, J. Gonzalez: Domain walls and magnetization reversal process in soft magnetic nanowires and nanotubes, J. Magn. Magn. Mat. **316**, 255 (2007)

[Vav06] P. Vavassori, R. Bovolenta, V. Metlushko, B. Ilic: Vortex rotation control in Permalloy disks with small circular voids, J. Appl. Phys. **99**, 053902 (2006)

[Váz04] M. Vázquez, M. Hernández-Vélez, K. Pirota, A. Asenjo, D. Navas, J. Velázquez, P. Vargas, and C. Ramos: Arrays of Ni nanowires in alumina membranes: magnetic properties and spatial ordering, Eur. Phys. J. B **40**, 489 (2004)

[Vaz05] C. A. F. Vaz, M. Kläui, L. J. Heyderman, C. David, F. Nolting, J. A. C. Bland: Multiplicity of magnetic domain states in circular elements probed by photoemission electron microscopy, Phys. Rev. B **72**, 224426 (2005)

[Ven12] K. S. Venkataraman, G. Dong, T. Zhang: Techniques Mitigating Update-Induced Latency Overhead in Shingled Magnetic Recording, IEEE Trans. Magn. **48**, 1899-1905 (2012)

[Vic05] R. H. Victora: Exchange Coupled Composite Media for Perpendicular Magnetic Recording, IEEE Transactions on Magnetics **41**, 537-542 (2005)

[Vic12] R. H. Victora, X. Chen: Exchange-assisted spin transfer torque switching, U. S. Patent 8,134,864 (March 13, 2012)

[Vil09] L. Vila, M. Darques, A. Encinas, U. Ebels, J.-M. George, G. Faini, A. Thiaville, and L. Piraux: Magnetic vortices in nanowires with transverse easy axis, Phys. Rev. B **79**, 172410 (2009)

[Wae06] B. Van Waeyenberge, A. Puzic, H. Stoll, K. W. Chou, T. Tyliszczak, R. Hertel, M. Fähnle, H. Brückl, K. Rott, G. Reiss, I. Neudecker, D. Weiss, C. H. Back & G. Schütz, Nature **444**, 461 (2006)

[Wan05] J. Wang, A. O. Adeyeye, and N. Singh: Magnetostatic interactions in mesoscopic Ni80Fe20 ring arrays, Appl. Phys. Lett. **87**, 262508 (2005)

[Wan09] R.-H. Wang, J.-S. Jiang, M. Hu: Metallic cobalt microcrystals with flowerlike architectures: Synthesis, growth mechanism and magnetic properties, Mater. Res. Bull. **44**, 1468 (2009)

[Wan12] W.-G. Wang, M. Li, S. Hagemann, and C. L. Chien: Electric-field-assisted switching in magnetic tunnel junctions, Nature Mater. **11**, 64 (2012)

[Wan13] W.-G. Wang and C. L. Chien: Voltage-induced switching in magnetic tunnel junctions with perpendicular magnetic anisotropy, J. Phys. D: Appl. Phys. **46**, 074004 (2013)

[Was07] R. Waser, M. Aono: Nanoionics-based Resistive Switching Memories, Nature Materials **6**, 833-840 (2007)

[Wel03] U. Welp, V. K. Vlasko-Vlasov, G. W. Crabtree, J. Hiller, N. Zaluzec, V. Metlushko, and B. Ilic: Magnetization reversal in arrays of individual and coupled Co-rings, J. Appl. Phys. **93**, 7056 (2003)

[Wel03a] U. Welp, V. K. Vlasko-Vlasov, J. M. Miller, N. J. Zaluzec, V. Metlushko, and B. Ilic: Magnetization reversal in arrays of Co rings, Phys. Rev. B **68**, 054408 (2003)

[Wes08] A. Westphalen, A. Schumann, A. Remhof, H. Zabel, M. Karolak, B. Baxevanis, E. Y. Vedmedenko, T. Last, U. Kunze, and T. Eimüller: Magnetization reversal of microstructured kagome lattices, Phys. Rev. B **77**, 174407 (2008)

[Wes13] WD VelociRaptor – Features (http://www.wdc.com/en/products/products.aspx?id=20)

[Woo11] K. Woojin, J. H. Jeong, Y. Kim, W. C. Lim, J.-H. Kim, J. H. Park, H. J. Shin, Y. S. Park, K. S. Kim, S. H. Park, Y. J. Lee, K. W. Kim,

H. J. Kwon, H. L. Park, H. S. Ahn, S. C. Oh, J. E. Lee, S. O. Park, S. Choi, H.-K Kang, C. Chung: Extended scalability of perpendicular STT-MRAM towards sub-20 nm MTJ node, Proc. of 2011 IEEE Int. Electron Devices Meeting (Washington, DC), pp. 24.1.1-14.1.4 (2011)

[Wri51] E. P. G. Wright: Electric connecting device, US Patent 2667542 (1951)

[Wu11] T. Wu, A. Bur, K. Wong, P. Zhao, Ch. S. Lynch, P. K. Amiri, K. L. Wang, and G. P. Carman: Electrical control of reversible and permanent magnetization reorientation for magnetoelectric memory devices, Appl. Phys. Lett 98, 262504 (2011)

[Yan11] J. K. W. Yang, Y. Chen, T. Huang, H. Duan, N. Thiyagarajah, H. K. Hui, S. H. Leong, and V. Ng: Fabrication and characterization of bit-patterned media beyond 1.5 Tbit/in², Nanotechnology 22, 385301 (2011)

[Yod10] H. Yoda, T. Kishi, T. Nagase, M. Yoshikawa, K. Nishiyama, E. Kitagawa, T. Daibou, M. Amano, N. Shimomura, and S. Takahashi: High efficient spin transfer torque writing on perpendicular magnetic tunnel junctions for high density MRAMs, Curr. Appl. Phys. 10, e87-89 (2010)

[Yoo12] S. Yoon, Y. Jang, K.-J. Kim, K.-W. Moon, J. Kim, Ch. Nam, S.-B. Choe, and B. K. Cho: Field and current induced asymmetric domain wall motion in a giant magnetoresistance spin-valve stripe with a circular ring, J. Appl. Phys. 111, 07B910 (2012)

[Zha10] W. Zhang and S. Haas: Phase diagram of magnetization reversal processes in nanorings, Phys. Rev. B 81, 064433 (2010)

[Zha10a] B. Zhang, W. Wang, C. Mu, Q. Liu, J. Wang: Calculations of three-dimensional magnetic excitations in permalloy nanostructures with vortex state, J. Magn. Magn. Mat. 322, 2480 (2010)

[Zho03] Z. Zhong, D. Wang, Y. Cui, M. W. Bockrath, C. M. Lieber: Nanowire Crossbar Arrays as Address Decoders for Integrated Nanosystems, Science 302, 1377 (2003)

[Zhu00] J. G. Zhu, Y. F. Zheng, and G. A. Prinz: Ultrahigh density vertical magnetoresistive random access memory (invited), J. Appl. Phys. 87, 6668 (2000)

[Zhu03] J.-G. Zhu: New heights for hard disk drives, meterialstoday July/August 2003, 22-30

[Zhu04] F. Q. Zhu, D. L. Fan, X. C. Zhu, J. G. Zhu, R. C. Cammarata, and C. L. Chien: Fabrication and magnetic properties of ultrahigh density arrays of ferromagnetic nanorings over larger areas, Adv. Mater. **16**, 2155 (2004)

[Zhu08] J.-G. Zhu, X. Zhu, and Y. Tang: Microwave assisted magnetic recording, IEEE Trans. Magn. **44**, 125-131 (2008)

Appendix I: Anisotropies and internal magnetic fields in FM systems

The search for magnetic systems with novel properties, which might be utilized in new magnetic storage systems, requires basic knowledge of magnetic anisotropies, which is given in this chapter.

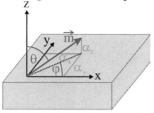

Fig. I.1: Coordinate system showing the spherical coordinates and the directional cosine terms for a thin-film sample.

Fig. I.1 depicts the spherical coordinates which will be used in the calculation of the phenomenological approach, with the azimuthal angle φ and the polar angle θ, the magnetization $\vec{m}$ and the directional cosine α:

$$\alpha_x = \cos \varphi \sin \theta$$
$$\alpha_y = \sin \varphi \sin \theta \qquad (I.1)$$
$$\alpha_z = \qquad \cos \theta$$

The sample surface, the x-y-plane, is, e.g., for (110) oriented samples identical with the crystallographic (110) plane, with the corresponding surface normal [110]. The angle between the x-direction and the direction of an external magnetic field in the sample plane is called φ_H. Next, some simple symmetry reasons can be used to find a phenomenological approach for the sample anisotropies. This procedure is illustrated here for a (110) oriented sample.

The magneto-crystalline anisotropy due to the crystal geometry can be derived from the magneto-crystalline energy

$$F_{mk} = K_1(\alpha_{x'}^2\alpha_{y'}^2 + \alpha_{y'}^2\alpha_{z'}^2 + \alpha_{z'}^2\alpha_{x'}^2), \qquad (I.2)$$

with the primed variables related to the (110) coordinate system. Transformation into the (100) system which is the crystallographic reference system can be done by a rotation of the lattice around the x-axis by $\pi/4$:

$$\begin{pmatrix} \alpha_{x'} \\ \alpha_{y'} \\ \alpha_{z'} \end{pmatrix} = \frac{1}{\sqrt{2}} \begin{pmatrix} \sqrt{2}\alpha_x \\ \alpha_y + \alpha_z \\ -\alpha_y + \alpha_z \end{pmatrix} \qquad (I.3)$$

This leads to the magneto-crystalline energy [Fas96, Hil90]

$$F_{mk} = \frac{K_1}{4}(2\alpha_x^2(\alpha_y + \alpha_z)^2 + (\alpha_y + \alpha_z)^2(-\alpha_y + \alpha_z)^2 + (-\alpha_y + \alpha_z)^2 \cdot 2\alpha_x^2)$$

$$= \frac{K_1}{4}\left(4\alpha_x^2\alpha_y^2 + 4\alpha_x^2\alpha_z^2 + \alpha_z^4 - 2\alpha_z^2\alpha_y^2 + \alpha_y^4\right) \tag{I.4}$$

$$= \frac{K_1}{4}(-3\sin^4\theta\sin^4\varphi + \sin^2\varphi(10\sin^4\theta - 6\sin^2\theta) + 2\sin^2\theta - 3\sin^4\theta + 1).$$

The second derivations according to θ and φ yield the internal anisotropy fields

$$H_{mk}^\alpha = \frac{1}{M_S}\frac{\partial^2 F_{mk}}{\partial\theta^2} \tag{I.5}$$

$$= \frac{K_1}{M_S}(-2 + 16\cos^2\theta - 14\cos^2\varphi\cos^2\theta + 3\cos^4\varphi + \cos^2\varphi$$
$$- 15\cos^4\varphi\cos^2\theta - 16\cos^4\theta + 16\cos^4\theta\cos^2\varphi + 12\cos^4\varphi\cos^4\theta)$$

and

$$H_{mk}^\beta = \frac{1}{M_S}\frac{\partial^2 F_{mk}}{\partial\varphi^2} \tag{I.6}$$

$$= \frac{K_1}{M_S}(-1 + \cos^2\theta)(11\cos^2\varphi - 12\cos^4\varphi - 5\cos^2\varphi\cos^2\theta$$
$$+ 12\cos^4\varphi\cos^2\theta - 2\cos^2\theta - 1).$$

If the magnetization is localized in the sample plane (i.e. $\theta = 90°$), these equations are simplified to

$$H_{mk}^\alpha = \frac{K_1}{M_S}(3\cos^4\varphi + \cos^2\varphi - 2) = \frac{K_1}{M_S}(3\sin^4\varphi - 7\sin^2\varphi + 2) \tag{I.7}$$

and

$$H_{mk}^\beta = \frac{K_1}{M_S}(12\cos^4\varphi - 11\cos^2\varphi + 1) = \frac{K_1}{M_S}(12\sin^4\varphi - 13\sin^2\varphi + 2). \tag{I.8}$$

Additionally, the magneto-elastic anisotropy has to be taken into account. It arises from tetragonal deformations in the layers due to lattice mismatches.

Such a deformation must include a two-fold symmetry, i.e. lead to a uniaxial anisotropy.

The phenomenological approach

$$F_{me} = -K_{out-of-plane}\,\alpha_z^2 + K_{in-plane}\,\alpha_{ref}^2$$

$$= -K_{out-of-plane}\cos^2\theta + K_{in-plane}\sin^2\theta\cos^2\Delta\varphi,$$

(I.9)

with the angle $\Delta\varphi = \varphi - \varphi_{ref}$ between magnetization and an easy crystal axis, leads to

$$H_{me}^{\alpha} = \frac{1}{M_S}\frac{\partial^2 F_{me}}{\partial\theta^2}$$

$$= \frac{1}{M_S}\frac{\partial}{\partial\theta}(-K_{out-of-plane}\cdot 2\cos\theta(-\sin\theta) + K_{in-plane}\cos^2\Delta\varphi\cdot 2\sin\theta\cos\theta)$$

(I.10)

$$= \frac{2}{M_S}(-K_{out-of-plane}(\sin^2\theta - \cos^2\theta) + K_{in-plane}\cos^2\Delta\varphi(\cos^2\theta - \sin^2\theta))$$

and with $\theta = 90°$ to

$$H_{me}^{\alpha} = \frac{2}{M_S}(-K_{out-of-plane} - K_{in-plane}\cos^2\Delta\varphi).$$

(I.11)

Similarly, the second derivation according to φ yields

$$H_{me}^{\beta} = \frac{1}{M_S}\frac{\partial^2 F_{me}}{\partial\varphi^2} = \frac{1}{M_S}\frac{\partial}{\partial\varphi}(K_{in-plane}\sin^2\theta\cdot 2\cos\Delta\varphi(-\sin\Delta\varphi))$$

$$= \frac{2}{M_S}K_{in-plane}\sin^2\theta(\sin^2\Delta\varphi - \cos^2\Delta\varphi) = \frac{2}{M_S}K_{in-plane}\sin^2\theta(2\sin^2\Delta\varphi - 1)$$

(I.12)

and for $\theta = 90°$

$$H_{me}^{\beta} = -\frac{2}{M_S}K_{in-plane}(1 - 2\sin^2\Delta\varphi).$$

(I.13)

Thus, a (110) oriented sample should contain fourfold and uniaxial (twofold) anisotropies, represented by terms of the forms $\sin^2\Delta\varphi$, $\cos^2\Delta\varphi$, and $\sin^4\Delta\varphi$.

Similarly, it can be shown that (100) oriented samples should contain a fourfold anisotropy, and a (111) oriented sample can be expected to show a sixfold anisotropy [Fas96, Hil90]. However, it has to be taken into account that nanostructured and other novel systems may show completely different anisotropies.

These formulas are used in the simulations depicted within this thesis and the interpretation of the experimental results.

Appendix II: Description of MathCad program used in this thesis

Opposite to MAGPAR, the program MathCad which is used in this chapter can handle a single macro-spin or a small ensemble of magnetic moments only. On the other hand, it offers the freedom to work with completely new forms of anisotropies, without any limitations by fixed models which are already included in a program. Thus, this chapter shows, starting from the dependence of coercivity and exchange bias on common anisotropies, the influence of the existence or absence of an intermediate state which can be explicitly "forbidden" in the mathematical descriptions in MathCad. Going further to novel anisotropies, different angular dependencies of the magnetic properties can be found, showing that such new anisotropy models are preferable in some magnetic systems.

Firstly, the MathCad program which has been written for the calculation of coercive fields and coercive peak heights will be explained. Colored areas are copied from the MathCad program.

$$\text{H} := 0..50 \qquad \phi_h := 0 \qquad\qquad \text{necessary starting parameters}$$

Firstly, starting parameters for the external field H and the angle of the field ϕ_h have to be defined, to simplify the definition of the calculated parameters later on. The values defined here are *not* used in the calculation.

$$M := \frac{21500}{4\pi} \qquad\qquad \text{material parameter}$$

As a material parameter, the saturation magnetization has to be defined, here for Fe. Oppositely to the micromagnetic simulations shown before, this parameter does not qualitatively influence the calculations performed in this chapter. Due to working with one single macro-spin now, the saturation magnetization only quantitatively influences the coercive fields which are linearly dependent on this parameter.

$$\text{Winkeldiff} := 5$$

$$\phi\_m := 0, \frac{\text{Winkeldiff} \,\pi}{180} ..2\pi$$

Simulation parameter: magnetization angle difference

The angular difference "Winkeldiff" is one of the simulation parameters, indicating the magnetization angle differences between subsequent calculations.

Physics parameters:
anisotropies [erg / cm^3]

$K_2 := 5000($ $K_4 := -0000($ $K_6 := 700($

The anisotropies are given in erg/cm³.

$$E(\phi\_m, H, \phi_h) := -H \cdot \cos(\phi_h) \cdot M \cdot \cos(\phi\_m) - H \cdot \sin(\phi_h) \cdot M \cdot \sin(\phi\_m) \dots$$
$$+ K_2 \cdot \sin\left(\phi\_m + 0\frac{\pi}{180}\right)^2 \dots$$
$$+ K_4 \cdot \cos\left(\phi\_m - 0.5\frac{\pi}{180}\right)^2 \cdot \sin\left(\phi\_m - 0.5\frac{\pi}{180}\right)^2 \dots$$
$$+ K_6 \cdot \cos\left(6\phi\_m + 120\frac{\pi}{180}\right)$$

Physics parameters in energy calculation: easy angles of anisotropies

The energy density is calculated taking into account the product **H M** as well as the anisotropies within the ferromagnet. The easy axes are defined by the angles given here. To avoid complications in calculations around 0° and other easy / hard axes of the anisotropies, it can be helpful to shift one of the axes slightly, as shown above.

$$\text{Steigung}(\phi\_m, H, \phi_h) := \frac{d}{d\phi\_m} E(\phi\_m, H, \phi_h)$$

calculation of energy derivation

The energy is differentiated to distinguish the slope of the energy landscape.

Energy landscape, depicted for some sample angles

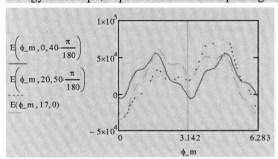

$$E\left(\phi\_m, 0, 40 \cdot \frac{\pi}{180}\right)$$

$$E\left(\phi\_m, 20, 50 \cdot \frac{\pi}{180}\right)$$

$$E(\phi\_m, 17, 0)$$

The energy landscape can be depicted for some angles, in order to verify that the anisotropies have been chosen in the correct way. Here, e.g., possible errors in the sign of the anisotropies can be found.

Derivative of energy, depicted for some sample angles

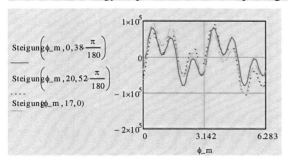

In the same way, the derivative of the energy is plotted for selected angles.

$$\varepsilon_{90°} := 90 \cdot \frac{\pi}{180}$$

$$K_{90°}(\phi\_m, H, \phi_h) := \begin{vmatrix} 1 & \text{if } \text{Steigung}(\phi\_m, H, \phi_h) > 0 \wedge \left[\left(\phi_h + \pi\right) \leq \phi\_m \leq \left(\phi_h + 2\pi - \varepsilon_{90°}\right) \vee \left[0 \leq \phi\_m \leq \left(\phi_h - \varepsilon_{90°}\right) \right] \right] \\ 2 & \text{if } \text{Steigung}(\phi\_m, H, \phi_h) < 0 \wedge \left[\left(\phi_h + \varepsilon_{90°}\right) \leq \phi\_m \leq \left(\phi_h + \pi\right) \right] \\ 0 & \text{otherwise} \end{vmatrix}$$

$$K_{90°\_1}(H, \phi_h) := \sum_{\phi\_m} \left[K_{90°}(\phi\_m, H, \phi_h) \cdot \left[\left(\phi_h + \varepsilon_{90°}\right) \leq \phi\_m < \left(\phi_h + \pi\right) \right] \right]$$

$$K_{90°\_2}(H, \phi_h) := \sum_{\phi\_m} \left[\left[K_{90°}(\phi\_m, H, \phi_h) \cdot \left[\left(\phi_h + \pi\right) < \phi\_m \leq \left(\phi_h + 2\pi - \varepsilon_{90°}\right) \right] \right] + \left[K_{90°}(\phi\_m, H, \phi_h) \cdot \left[0 \leq \phi\_m < \left(\phi_h - \varepsilon_{90°}\right) \right] \right] \right]$$

The calculation of the coercive fields $K_{90°}$ starts as follows: Depending on the energy derivative, the system decides in which direction (clockwise or counter-clockwise) the macrospin starts to rotate when the external magnetic field is reduced (possibilities 1 and 2). As soon as the energy density decreases continuously from positive saturation magnetization to a state perpendicular to the external field, the longitudinal magnetization component vanishes, i.e. the coercive field is reached.

$n := 0, 2.. 180$ max. 180°!

Simulation parameter:
distance between simulated field angles

$$\text{Koerz}_{90°}(n) := \text{ for } \phi_h \in n \cdot \frac{\pi}{180}$$

$$\begin{array}{|l} H \leftarrow 20 \\ \text{while } K_{90°\_1}(H, \phi_h) > 0 \wedge K_{90°\_2}(H, \phi_h) > 0 \\ \quad H \leftarrow H + 5 \\ H \end{array}$$

This procedure is carried out for the angles 0°, n°, 2n° ... 180°. In the easiest case, a brute force method is used to find the coercive field, by fitting the initial field H and the steps in which H should be increased (here by 5). A more advanced calculation using nested intervals can enhance the calculation speed (not shown here).

Calculated coercivities – graph and numbers for export to Origin / Excel

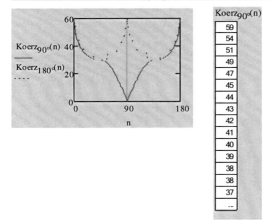

$\text{Koerz}_{90°}(n)$
59
54
51
49
47
45
44
43
42
41
40
39
38
38
37
...

As a result, the coercivities can be depicted in the form of a graph and additionally as numbers which can be exported to Excel or Origin.

Appendix III: Comparison of theoretical and experimental results

Most of the experimental and micromagnetic simulated results could be reproduced by a macrospin model used for calculations in MathCad. Table III.1 gives an overview of the formula used together with the corresponding results from experiments and micromagnetic simulations.

Table III.1

MathCad terms	Magpar simulation	Experimental sample
$\sin^2(\phi_m)\cos^2(\phi_m)$	2 pairs of wires without coupling between pairs	MgO(100)/Co/CoO
$\sin^2(\phi_m)$	1 pair of wires	
$\sin^2(\phi_m)$, $\sin^2(\phi_m)$ $\cos^2(\phi_m)$	Rhombic wire system	
$\sin^2(\phi_m)$ $\cos^2(\phi_m)$, $|\sin^2(\phi_m)|$, Gaussian distribution		
$\sum|\sin^2(\phi_m + n\cdot60°)|$	Sixfold wire samples	